Collins

AF513245

Collins
Science Workbook

6

Published by Collins Learning
A division of HarperCollins*Publishers* India Private Limited

HarperCollins*Publishers*

A-75, Sector-57, Noida, Uttar Pradesh 201301, India

1 London Bridge Street, London, SE1 9GF, United Kingdom

2 Bloor Street East, Toronto, Ontario M4W 1A8, Canada

Lvl 13, 201 Elizabeth Street (PO Box A565, NSW, 1235), Sydney NSW 2000, Australia

195 Broadway, New York, NY 10007, USA

Browse the complete Collins catalogue at **www.collins.in**

For any queries/feedback, please write to us at
SCHOOLSUPPORT@harpercollins.co.in or **call us on 0120-4044800.**

First edition 2020

Reprint: 10 9 8 7 6 5 4 3 2 1

ISBN: 978-93-5357-163-4

Content developed by Excelsoft Technologies Pvt. Ltd.

Artwork and layout by AptaraCorp Ltd., Mammoth Studioz Ltd.

Typesetting design and layout by Interactive Book Services Pvt. Ltd.

Cover design by Interactive Book Services Pvt. Ltd.

Printed and bound in India by MicroPrints India, New Delhi

Acknowledgements

The publisher wishes to thank the following for permission to use copyrighted images.

Cover – Luvi 40

Inside – Page **10:** highviews, diamant24; **13:** Kantawan, Mogens Trolle; **15:** davemhuntphotography, BOONROONG; **19:** Nata-Lia, wawri, Robyn Mackenzie, daffodilred, Madlen; **24:** sta, studioVin; **28:** discpicture; **31:** Xiebiyun; **49:** iofoto; **51:** ifong; **55:** Anan Kaewkhammul, farmer79; **58:** Myibean, farmer79; **73:** Copit, tchara; **78:** think4photop, MR. SUWIT GAE, Madlen; **94:** Jarry, aleksandr hunta; **112:** Reshavskyi; **118:** magicoven, ktynzq, Fedorov Oleksiy, PeJo, Bedrin; **139:** Sunny Forest; **145:** Aksenova Natalya; **154:** vilax; **159:** Svend77, wawritto

Preface

Collins Science Workbook is a series of three workbooks for classes 6 to 8, which are aimed at providing comprehensive practice for the development and understanding of scientific concepts. The workbooks are in accordance with the recommendations of the National Curriculum Framework 2005 and the syllabus prescribed by National Council of Educational Research and Training (NCERT).

As practice and review is one of the most important part of learning and strengthening the child's foundation in any subject, the worksheets in the workbooks have been designed to help develop essential understanding in key scientific concepts. The multifaceted worksheets (five worksheets per chapter) are planned in such a way that they become progressively more challenging to ensure complete understanding and application of the concepts.

The workbooks can be used by teachers for the purpose of review and assessments, if they want to, as the worksheets have been designed that way, incorporating a variety of questions.

Key Features

Recap of the chapter	Main concepts of the chapter for quick recall
Concept Map	Summary of the chapter in the form of a flowchart
Worksheets	Five worksheets per chapter
Worksheet 1	Objective type questions based on the first half of the chapter
Worksheet 2	Short answer type questions based on the first half of the chapter
Worksheet 3	Objective type questions based on the second half of the chapter
Worksheet 4	Short answer type questions based on the second half of the chapter
Worksheet 5	Objective and subjective type questions based on the entire chapter

We would like to take this opportunity to thank all the teachers who reviewed the workbooks and provided their feedback, which helped in improving the quality of the content.

Feedback, invaluable comments and suggestions from users are welcome.

Authors

Key Features

Each chapter comprises five multifaceted worksheets. First half of the chapter is covered in Worksheet 1 and Worksheet 2. The second half of the chapter is covered in Worksheet 3 and Worksheet 4. The complete chapter is covered in Worksheet 5.

Worksheets 1, 2, 3 and 4 can be used as a short review and assessment of **key topics**. Worksheet 5 provides a comprehensive approach and contains questions of all the topics of the chapter to reinforce problem-solving skills.

Marking Schemes

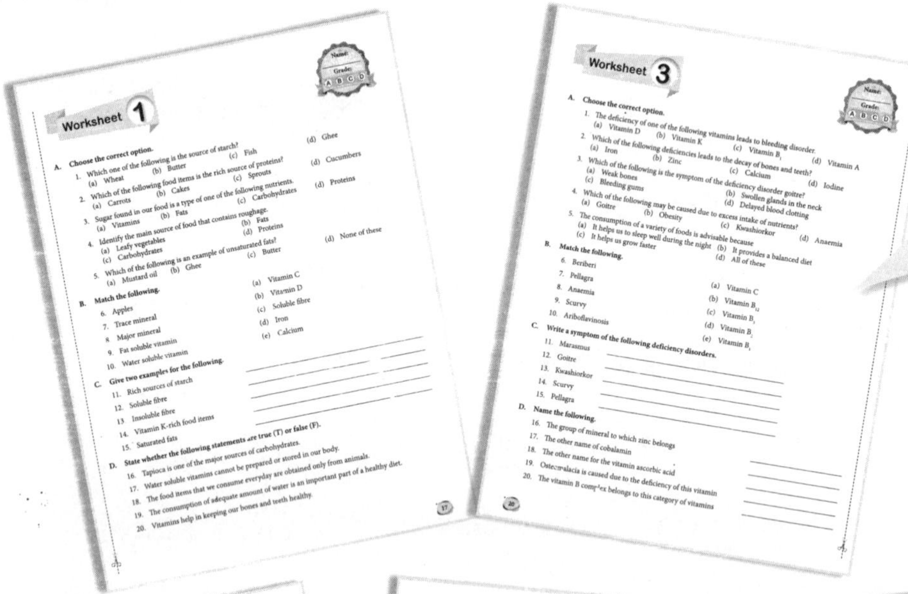

Each question in Worksheet 1 and Worksheet 3 carries 1 mark.

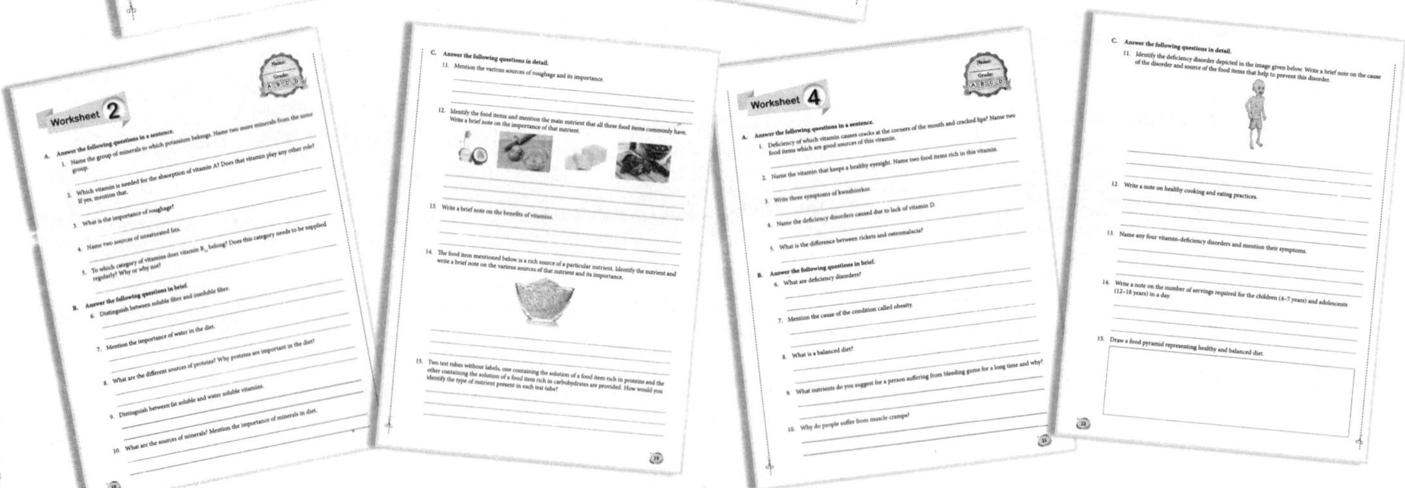

Worksheet 2 and Worksheet 4 comprise of three sections A, B and C (with 5 questions each). Each question in sections A, B and C carries 1, 2 and 3 marks respectively.

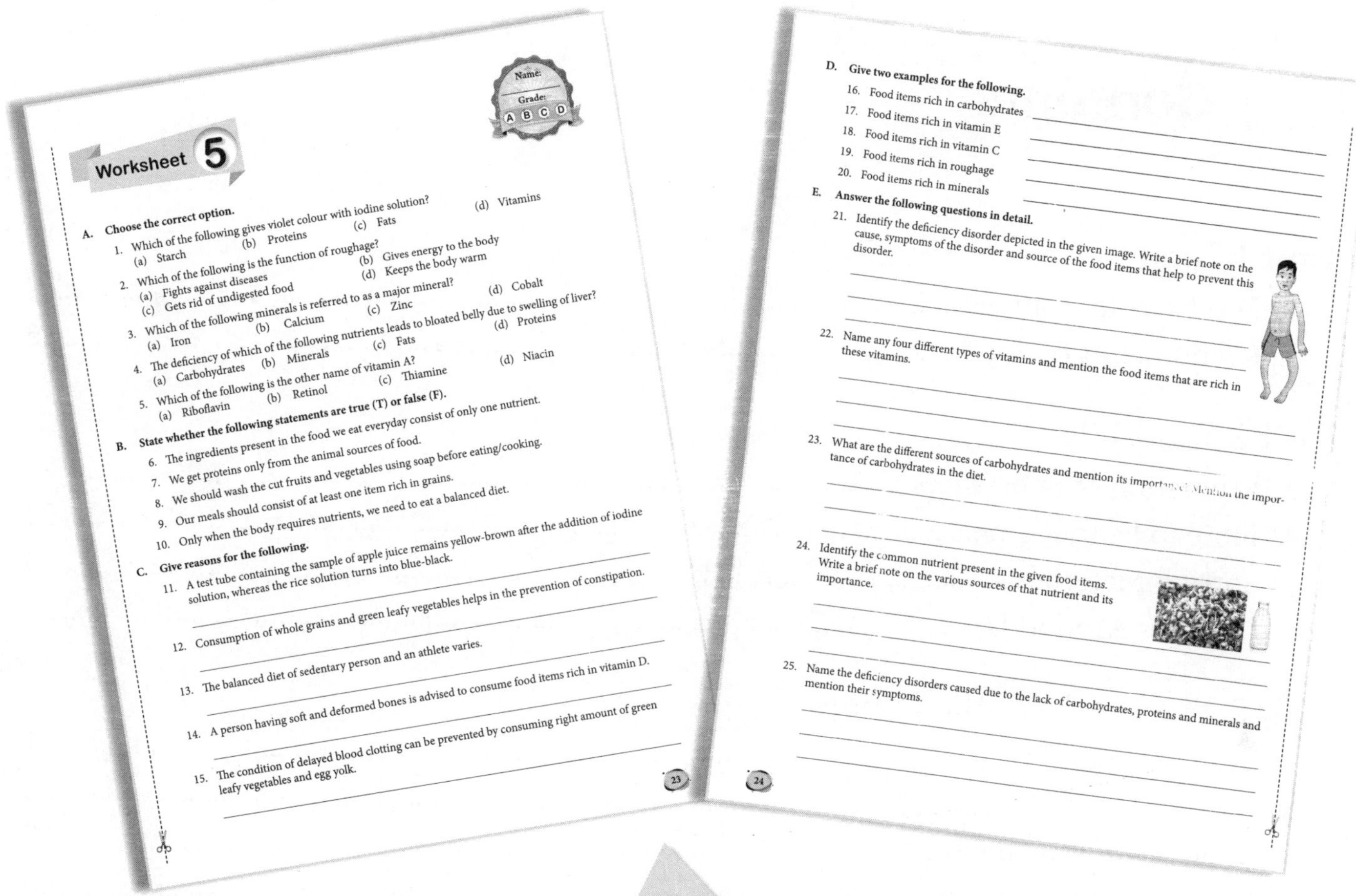

Name:

Grade: A B C D

Worksheet 5

A. Choose the correct option.

1. Which of the following gives violet colour with iodine solution?
(a) Starch (b) Proteins (c) Fats (d) Vitamins
2. Which of the following is the function of roughage?
(a) Fights against diseases (b) Gives energy to the body
(c) Gets rid of undigested food (d) Keeps the body warm
3. Which of the following minerals is referred to as a major mineral?
(a) Iron (b) Calcium (c) Zinc (d) Cobalt
4. The deficiency of which of the following nutrients leads to bloated belly due to swelling of liver?
(a) Carbohydrates (b) Minerals (c) Fats (d) Proteins
5. Which of the following is the other name of vitamin A?
(a) Riboflavin (b) Retinol (c) Thiamine (d) Niacin

B. State whether the following statements are true (T) or false (F).

6. The ingredients present in the food we eat everyday consist of only one nutrient.
7. We get proteins only from the animal sources of food.
8. We should wash the cut fruits and vegetables using soap before eating/cooking.
9. Our meals should consist of at least one item rich in grains.
10. Only when the body requires nutrients, we need to eat a balanced diet.

C. Give reasons for the following.

11. A test tube containing the sample of apple juice remains yellow-brown after the addition of iodine solution, whereas the rice solution turns into blue-black.
12. Consumption of whole grains and green leafy vegetables helps in the prevention of constipation.
13. The balanced diet of sedentary person and an athlete varies.
14. A person having soft and deformed bones is advised to consume food items rich in vitamin D.
15. The condition of delayed blood clotting can be prevented by consuming right amount of green leafy vegetables and egg yolk.

23

D. Give two examples for the following.

16. Food items rich in carbohydrates
17. Food items rich in vitamin E
18. Food items rich in vitamin C
19. Food items rich in roughage
20. Food items rich in minerals

E. Answer the following questions in detail.

21. Identify the deficiency disorder depicted in the given image. Write a brief note on the cause, symptoms of the disorder and source of the food items that help to prevent this disorder.
22. Name any four different types of vitamins and mention the food items that are rich in these vitamins.
23. What are the different sources of carbohydrates and mention its importance? Mention the importance of carbohydrates in the diet.
24. Identify the common nutrient present in the given food items. Write a brief note on the various sources of that nutrient and its importance.
25. Name the deficiency disorders caused due to the lack of carbohydrates, proteins and minerals and mention their symptoms.

24

In Workbook 6, Worksheet 5 comprises of five sections: A, B, C, D and E (with 5 questions each). Each question in sections A, B, C, D and E carries 1, 1, 2, 2 and 5 marks respectively.

In Workbooks 7 and 8, Worksheet 5 comprises of seven sections: A, B, C, D, E, F and G (with 5 questions each). Each question in sections A, B, C, D, E, F and G carries 1, 1, 2, 2, 1, 2 and 5 marks respectively.

Suggestive Grading System	
Marks %	**Grade**
81–100	A
61–80	B
41–60	C
33–40	D

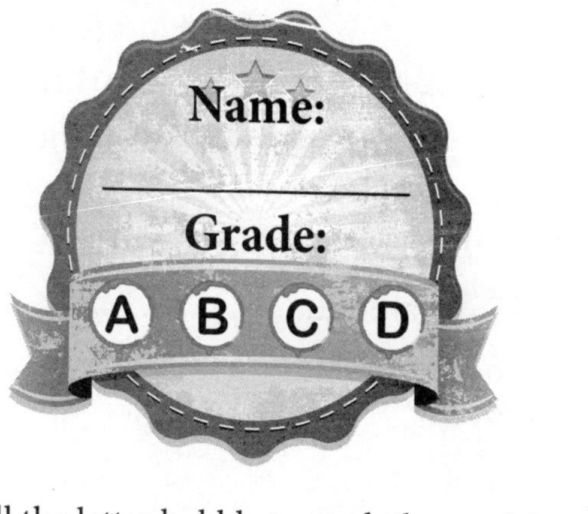

Fill the letter bubble to mark the grade!

Contents

1 Food and Its Sources

Recap of the Chapter

- Nutrients present in food help us to grow and stay healthy.
- Nutrients can be obtained from plant as well as animal sources of food.
- A major portion of the food being consumed is called staple food.
- Animals are grouped into various categories like herbivores, carnivores and omnivores based on their food habits.
- Animals that eat dead remains of other animals are called scavengers.
- Animals that feed on other animals of their own kind are called cannibals.
- Living organisms that feed on dead and decaying remains of animals and plants are called decomposers.
- Living organisms that live in or on another living thing and depend on it for food are called parasites.

Concept Map

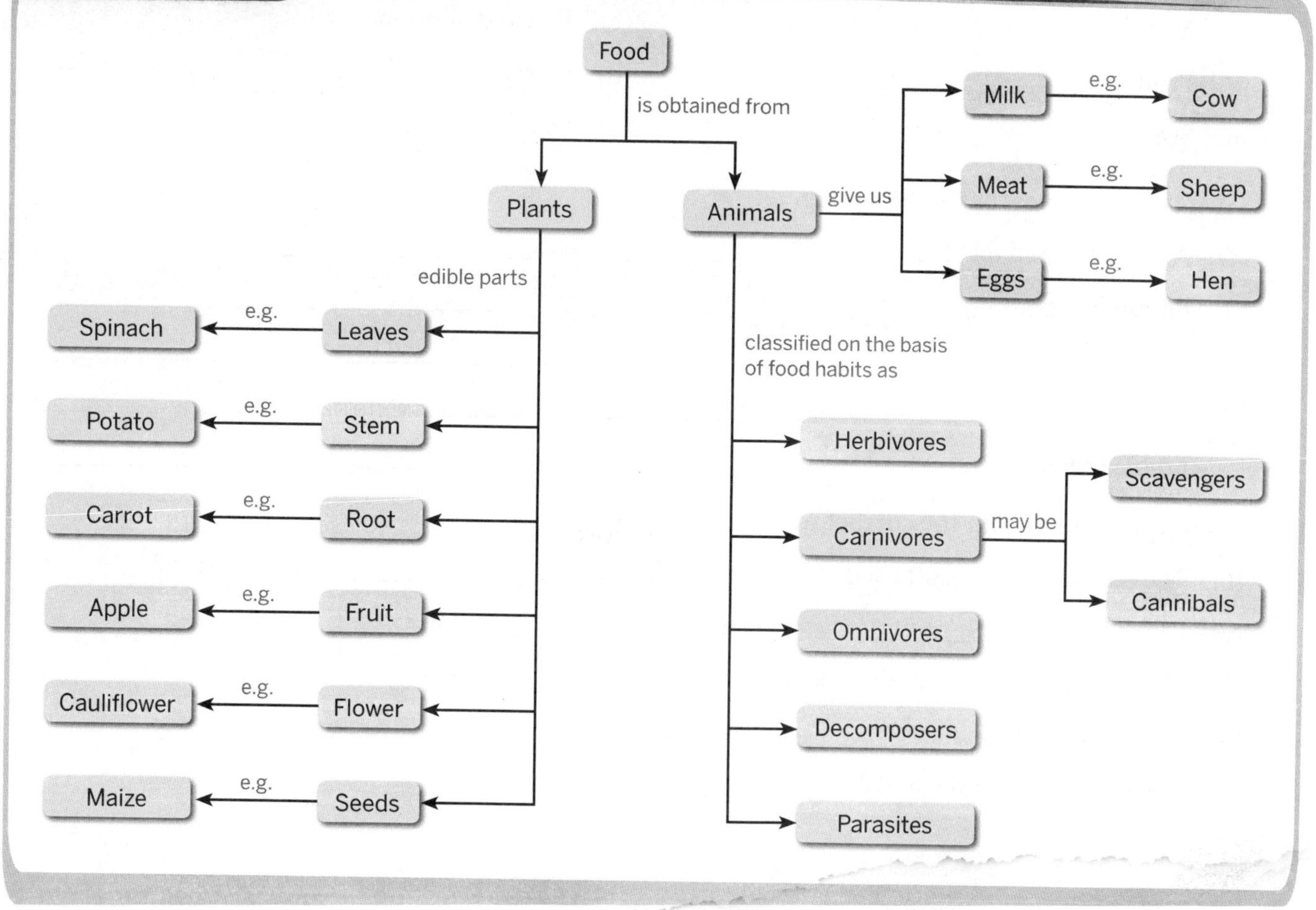

Worksheet 1

Name:

Grade: A B C D

A. Choose the correct option.

1. Which of the following is a dairy product?
   (a) Butter (b) Sunflower oil (c) Duck's egg (d) Honey
2. Which of the following bird's egg is consumed?
   (a) Sparrow (b) Hummingbird (c) Duck (d) Peahen
3. Which of the following animals is reared for its meat?
   (a) Sheep (b) Horse (c) Donkey (d) Dog
4. Which of the following seeds is used to make food items?
   (a) Lemon seed (b) Apple seed (c) Mango seed (d) Maize
5. Which of the following is an edible flower?
   (a) Apple (b) Carrot (c) Spinach (d) Broccoli

B. Match the following.

6.	Wheat	(a)	*Kheer*
7.	Milk	(b)	Chapati
8.	Meat	(c)	Roasted almonds
9.	Fruit	(d)	Curry
10.	Seeds	(e)	Salad

C. Give two examples for the following.

11. Animals reared for their eggs ____________________
12. Edible flowers ____________________
13. Edible leaves ____________________
14. Edible fruits ____________________
15. Edible seeds ____________________

D. Fill in the blanks using the correct words.

16. The part of a mint plant used to prepare medicine is its ____________. (roots/leaves)
17. Clove is a dried ____________. (flower/stem)
18. Tender ____________ of the celery plant are edible. (stalk/root)
19. Turmeric is a modified ____________. (bud/stem)
20. Turnip is an example of edible ____________. (flower/root)

Worksheet 2

Name:

Grade: A B C D

A. Answer the following questions in a sentence.

1. Name the parts of potato, pea plant and spinach which are consumed as food.

2. Name three plant parts that are used as spice?

3. Name any three pulses used in Indian cuisine.

4. Name three plants, roots of which are eaten as food.

5. Define staple food.

B. Answer the following questions in brief.

6. Describe the uses of seeds as ingredients in food items.

7. Which part of bamboo is used as food? Write one benefit of eating this part.

8. Which parts of hen are used as food? Name two dishes prepared with them.

9. Which plant leaves are used in garnishing food? What are the other uses of these leaves?

10. What is meant by dairy products? Give examples.

C. Answer the following questions in detail.

11. Explain various animal sources of food with examples.

____

____

____

12. Explain the process of extracting honey.

____

____

____

13. Categorize the given sources of food into animal sources and plant sources.
(Rice flour, milk, cheese, cinnamon, mustard oil and frozen fish)

Plant source	Animal source

14. Name the following food items, the ingredients used to prepare them and their source (plant or animal).

Name ______________________

Ingredients ______________________

Source ______________________

Name ______________________

Ingredients ______________________

Source ______________________

15. Name and identify the sources of ingredients from Anna's lunch.

Ingredients ______________________

Sources ______________________

Worksheet 3

Name: ____________

Grade: Ⓐ Ⓑ Ⓒ Ⓓ

A. Choose the correct option.

1. Which of the following is a carnivore?
   (a) Caterpillar (b) Cow (c) Crocodile (d) Camel
2. Which of the following birds is a herbivore?
   (a) Sparrow (b) Crow (c) Duck (d) Parakeet
3. Which of the following animals has proboscis?
   (a) Honeybee (b) Butterfly (c) Mosquito (d) All of these
4. Which of the following animals eats both plants and other animals?
   (a) Crow (b) Buffalo (c) Hippopotamus (d) Giraffe
5. Which of the following is an example of cannibal?
   (a) Mosquito (b) Dragonfly (c) Housefly (d) Cockroach

B. State whether the following statements are true (T) or false (F).

6. Herbivores eat only the meat of other animals.
7. Animals that eat both plants and other animals are called cannibals.
8. Spider is an example of a parasite.
9. Parasites feed on dead and decaying matter.
10. Hyena is an example of an herbivore.

C. Name the following.

11. Animals that feed only on plant parts ____________
12. Animals that eat only the meat of other animals ____________
13. Organisms that feed on dead and decaying plants or animals ____________
14. Animals that eat their own kind ____________
15. Organisms that live on or inside another animal ____________

D. Fill in the blanks using the correct words.

16. Rat is an example of ____________ animal. (herbivorous/omnivorous)
17. Frog feeds on insects using its ____________. (leg/tongue)
18. Leech depends on other ____________ organisms for food. (living/dead)
19. Dead remains of animals are eaten by ____________. (herbivores/scavengers)
20. Tapeworm is an example of ____________. (parasites/cannibals)

Worksheet 4

Name: ____________

Grade: (A) (B) (C) (D)

A. Answer the following questions in a sentence.

1. Name any two organisms that are decomposers.

   ____

2. Why do spiders feed on their own kind?

   ____

3. To which category of animals does the bear belong according to its food habit? Name two more animals that belong to the same category.

   ____

4. Give two examples of the birds that are carnivorous.

   ____

5. Which body part is common in butterfly and honeybee? What is its function?

   ____

B. Answer the following questions in brief.

6. Mention all the categories of animals based on their food habits.

   ____

   ____

7. Which parts of the mouth help carnivores eat the meat of other animals and how?

   ____

   ____

8. Describe the food habits of an omnivore.

   ____

   ____

9. How is the food habit of a crow and vulture similar?

   ____

   ____

10. What is common between carnivorous and herbivorous animals? What is the major difference between the two?

    ____

    ____

C. **Answer the following questions in detail.**

11. Describe the structure of teeth in plant-eating animals.

12. Differentiate between a parasite and a decomposer with respect to their food habits.

13. Describe the feeding habit of an animal similar to that of a frog?

14. Mosquitoes are not cannibals. Justify the given statement.

15. Identify the body part and write its function that helps the given animals to consume their food.

(a) 

(b) 

(c) 

Worksheet 5

Name:

Grade: A B C D

A. Choose the correct option.

1. Which of the following is an omnivore?
   (a) Horse (b) Hedgehog (c) Frog (d) Spider
2. Which of the following is a parasite?
   (a) Mosquito (b) Dog (c) Dragonfly (d) Lion
3. Which of the following is a herbivore?
   (a) Parakeet (b) Sparrow (c) Panda (d) Vulture
4. Which of the following lives on or within the body of another organism?
   (a) Earthworm (b) Caterpillar (c) Tapeworm (d) Centipede
5. Which of the following is an edible stem?
   (a) Sugarcane (b) Ginger (c) Turmeric (d) All of these

B. Match the following.

6. Stem	(a) Beetroot
7. Root	(b) Mango
8. Flower	(c) Potato
9. Seeds	(d) Cauliflower
10. Fruit	(e) Maize

C. Give reasons for the following.

11. Scavengers are also called carnivores.

____

12. Carnivores have sharp and pointed teeth.

____

13. Honey is also used as a natural sweetener.

____

14. Stalks of celery and bamboo are edible.

____

15. Different animals have different food habits.

____

D. Complete the pattern.

16.	Butterfly	:	Nectar	::	Lion	: ________
17.	Horse	:	________	::	Mosquito	: Blood
18.	Buffalo	:	Milk	::	________	: Honey
19.	Cow	:	Herbivore	::	Tiger	: ________
20.	Crow	:	Scavenger	::	Mosquito	: ________

E. Answer the following questions in detail.

21. Write the similarities between decomposing microorganisms and animal scavengers.

22. Name the edible parts of plants and give one example for each.

23. List the reasons for shortage of food and the appropriate measures taken to prevent it.

24. Do birds and insects damage the flowers while sucking the nectar? Justify your answer.

25. Categorize the given birds based on their food habits as herbivores, carnivores or omnivores. Describe their food habit.

(a) 

Category ________

Food habit ________

(b) 

Category ________

Food habit ________

2 Components of Food

Recap of the Chapter

- The food we eat contains nutrients that are required for our body to grow and stay healthy.
- The major nutrients required by our body are carbohydrates, proteins, fats, vitamins and minerals in addition to dietary fibre and water.
- Carbohydrates are the main source of energy and form a major part of our staple diet.
- Proteins are the building blocks of the human body.
- There are two types of vitamins—fat soluble and water soluble, that help us to develop immunity against many diseases.
- Minerals are required for making our bones strong, building muscles and proper functioning of the body.
- Roughage is called dietary fibre. It helps in getting rid of undigested food and waste from the body.
- Balanced diet contains all the nutrients in right proportions.
- Deficiency of nutrients causes deficiency diseases while excess intake of nutrients causes obesity.

Concept Map

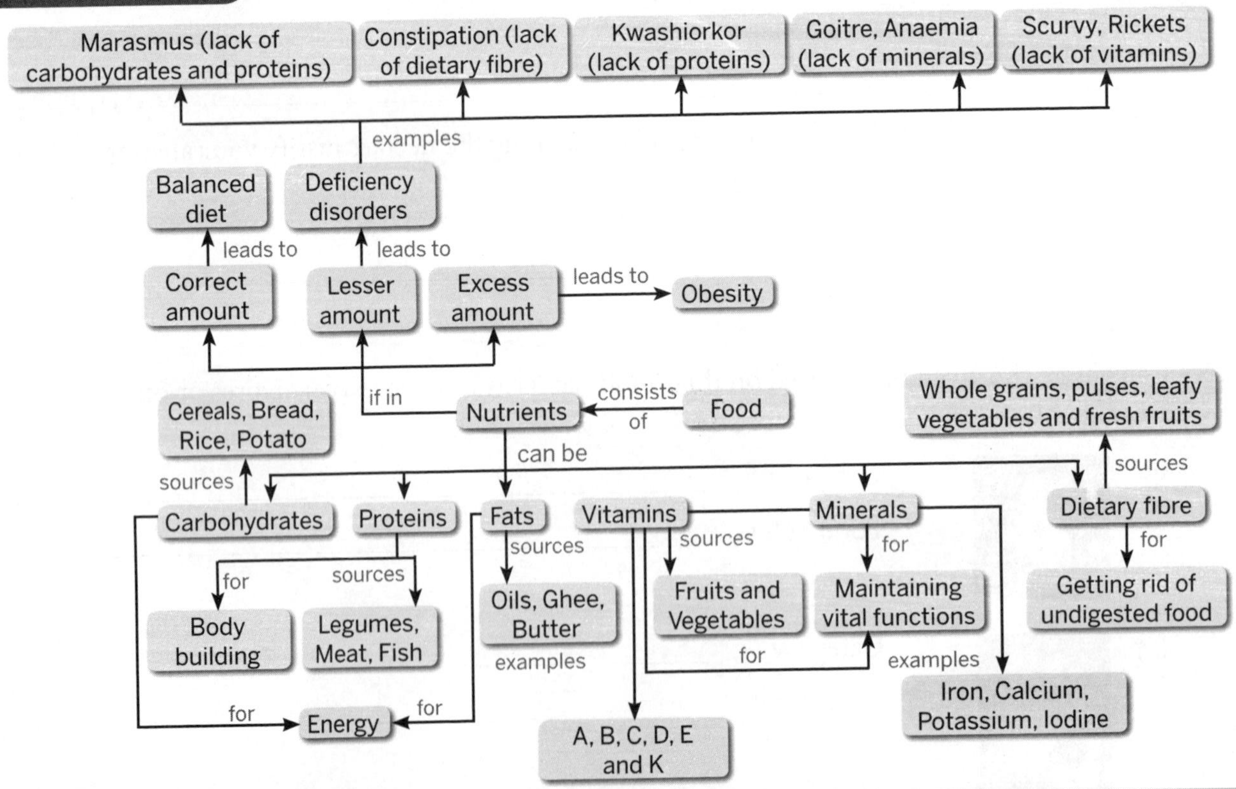

Worksheet 1

Name:

Grade: A B C D

A. Choose the correct option.

1. Which one of the following is the source of starch?
   (a) Wheat (b) Butter (c) Fish (d) Ghee
2. Which of the following food items is the rich source of proteins?
   (a) Carrots (b) Cakes (c) Sprouts (d) Cucumbers
3. Sugar found in our food is a type of one of the following nutrients.
   (a) Vitamins (b) Fats (c) Carbohydrates (d) Proteins
4. Identify the main source of food that contains roughage.
   (a) Leafy vegetables (b) Fats
   (c) Carbohydrates (d) Proteins
5. Which of the following is an example of unsaturated fats?
   (a) Mustard oil (b) Ghee (c) Butter (d) None of these

B. Match the following.

6. Apples	(a) Vitamin C
7. Trace mineral	(b) Vitamin D
8. Major mineral	(c) Soluble fibre
9. Fat soluble vitamin	(d) Iron
10. Water soluble vitamin	(e) Calcium

C. Give two examples for the following.

11. Rich sources of starch ____________________
12. Soluble fibre ____________________
13. Insoluble fibre ____________________
14. Vitamin K-rich food items ____________________
15. Saturated fats ____________________

D. State whether the following statements are true (T) or false (F).

16. Tapioca is one of the major sources of carbohydrates.
17. Water soluble vitamins cannot be prepared or stored in our body.
18. The food items that we consume everyday are obtained only from animals.
19. The consumption of adequate amount of water is an important part of a healthy diet.
20. Vitamins help in keeping our bones and teeth healthy.

Worksheet 2

Name: ______

Grade: A B C D

A. Answer the following questions in a sentence.

1. Name the group of minerals to which potassium belongs. Name two more minerals from the same group.

____

2. Which vitamin is needed for the absorption of vitamin A? Does that vitamin play any other role? If yes, mention that.

____

3. What is the importance of roughage?

____

4. Name two sources of unsaturated fats.

____

5. To which category of vitamins does vitamin B_{12} belong? Does this category needs to be supplied regularly? Why or why not?

____

B. Answer the following questions in brief.

6. Distinguish between soluble fibre and insoluble fibre.

____

____

7. Mention the importance of water in the diet.

____

____

8. What are the different sources of proteins? Why proteins are important in the diet?

____

____

9. Distinguish between fat-soluble and water-soluble vitamins.

____

____

10. What are the sources of minerals? Mention the importance of minerals in a diet.

____

____

C. Answer the following questions in detail.

11. Mention the various sources of roughage and its importance.

12. Identify the food items and mention the main nutrient that all these food items commonly have. Write a brief note on the importance of that nutrient.

13. Write a brief note on the benefits of vitamins.

14. The food item mentioned below is a rich source of a particular nutrient. Identify the nutrient and write a brief note on the various sources of that nutrient and its importance.

15. Two test tubes without labels, one containing the solution of a food item rich in proteins and the other containing the solution of a food item rich in carbohydrates, are provided. How would you identify the type of nutrient present in each test tube?

Worksheet 3

Name:

Grade: A B C D

A. Choose the correct option.

1. The deficiency of one of the following vitamins leads to bleeding disorder.
   (a) Vitamin D (b) Vitamin K (c) Vitamin B_1 (d) Vitamin A
2. Which of the following deficiencies leads to the decay of bones and teeth?
   (a) Iron (b) Zinc (c) Calcium (d) Iodine
3. Which of the following is the symptom of the deficiency disorder goitre?
   (a) Weak bones (b) Swollen glands in the neck
   (c) Bleeding gums (d) Delayed blood clotting
4. Which of the following may be caused due to excess intake of nutrients?
   (a) Goitre (b) Obesity (c) Kwashiorkor (d) Anaemia
5. The consumption of a variety of foods is advisable because
   (a) It helps us to sleep well during the night (b) It provides a balanced diet
   (c) It helps us grow faster (d) All of these

B. Match the following.

6. Beriberi	(a) Vitamin C
7. Pellagra	(b) Vitamin B_{12}
8. Anaemia	(c) Vitamin B_3
9. Scurvy	(d) Vitamin B_2
10. Ariboflavinosis	(e) Vitamin B_1

C. Write a symptom of the following deficiency disorders.

11. Marasmus ________________
12. Goitre ________________
13. Kwashiorkor ________________
14. Scurvy ________________
15. Pellagra ________________

D. Name the following.

16. The group of mineral to which zinc belongs ________________
17. The other name of cobalamin ________________
18. The other name for the vitamin ascorbic acid ________________
19. Osteomalacia is caused due to the deficiency of this vitamin ________________
20. The vitamin B complex belongs to this category of vitamins ________________

Worksheet 4

Name:

Grade: A B C D

A. Answer the following questions in a sentence.

1. Deficiency of which vitamin causes cracks at the corners of the mouth and cracked lips? Name two food items which are good sources of this vitamin.

2. Name the vitamin that keeps a healthy eyesight. Name two food items rich in this vitamin.

3. Write three symptoms of kwashiorkor.

4. Name the deficiency disorders caused due to lack of vitamin D.

5. What is the difference between rickets and osteomalacia?

B. Answer the following questions in brief.

6. What are deficiency disorders?

7. Mention the cause of the condition called obesity.

8. What is a balanced diet?

9. What nutrients do you suggest for a person suffering from bleeding gums for a long time and why?

10. Why do people suffer from muscle cramps?

C. Answer the following questions in detail.

11. Identify the deficiency disorder depicted in the image given below. Write a brief note on the cause of the disorder and source of the food items that help to prevent this disorder.

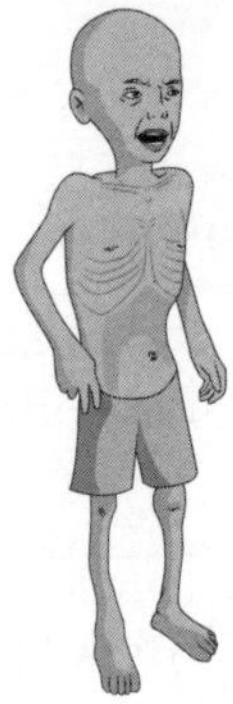

12. Write a note on healthy cooking and eating practices.

13. Name any four vitamin-deficiency disorders and mention their symptoms.

14. Write a note on the number of servings required for the children (4–7 years) and adolescents (12–18 years) in a day.

15. Draw a food pyramid representing healthy and balanced diet.

Worksheet 5

Name:

Grade: A B C D

A. Choose the correct option.

1. Which of the following gives blue-black colour with iodine solution?
   (a) Starch (b) Proteins (c) Fats (d) Vitamins
2. Which of the following is the function of roughage?
   (a) Fights against diseases (b) Gives energy to the body
   (c) Gets rid of undigested food (d) Keeps the body warm
3. Which of the following minerals is referred to as a major mineral?
   (a) Iron (b) Calcium (c) Zinc (d) Cobalt
4. The deficiency of which of the following nutrients leads to bloated belly due to swelling of liver?
   (a) Carbohydrates (b) Minerals (c) Fats (d) Proteins
5. Which of the following is the other name of vitamin A?
   (a) Riboflavin (b) Retinol (c) Thiamine (d) Niacin

B. State whether the following statements are true (T) or false (F).

6. The ingredients present in the food we eat everyday consist of only one nutrient.
7. We get proteins only from the animal sources of food.
8. We should wash the cut fruits and vegetables using soap before eating/cooking.
9. Our meals should consist of at least one item rich in grains.
10. Only when the body requires nutrients, we need to eat a balanced diet.

C. Give reasons for the following.

11. A test tube containing the sample of apple juice remains yellow-brown after the addition of iodine solution, whereas the rice solution turns into blue-black.

    ____

12. Consumption of whole grains and green leafy vegetables helps in the prevention of constipation.

    ____

13. The balanced diet of sedentary person and an athlete varies.

    ____

14. A person having soft and deformed bones is advised to consume food items rich in vitamin D.

    ____

15. The condition of delayed blood clotting can be prevented by consuming right amount of green leafy vegetables and egg yolk.

    ____

D. **Give two examples for the following.**

16. Food items rich in carbohydrates ____________________
17. Food items rich in vitamin E ____________________
18. Food items rich in vitamin C ____________________
19. Food items rich in roughage ____________________
20. Food items rich in minerals ____________________

E. **Answer the following questions in detail.**

21. Identify the deficiency disorder depicted in the given image. Write a brief note on the cause, symptoms of the disorder and source of the food items that help to prevent this disorder.

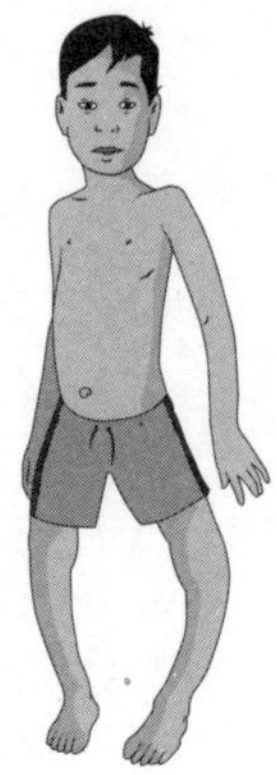

____

____

____

22. Name any four different types of vitamins and mention the food items that are rich in these vitamins.

____

____

____

23. What are the different sources of carbohydrates? Mention the importance of carbohydrates in the diet.

____

____

____

24. Identify the common nutrient present in the given food items. Write a brief note on the various sources of that nutrient and its importance.

____

____

____

25. Name the deficiency disorders caused due to the lack of carbohydrates, proteins and minerals and mention their symptoms.

____

____

____

3 Separation of Substances

Recap of the Chapter

- Substances can either be pure or mixture.
- Based on the state of components mixed together, mixtures can be classified into solid–solid mixture, solid–liquid mixture and liquid–liquid mixture.
- Based on the extent to which two or more components can be mixed, mixtures can be classified as homogeneous and heterogeneous.
- Separation of mixtures into their constituents is necessary to separate useful components from a mixture, to remove impurities and to get desirable substances.
- Different methods of separation include separation of solid-solid mixtures and solid-liquid mixtures.
- Separation of solid-solid mixtures—handpicking, threshing, winnowing and sieving.
- Separation of solid-liquid mixtures—sedimentation and decantation, filtration, evaporation and condensation.

Concept Map

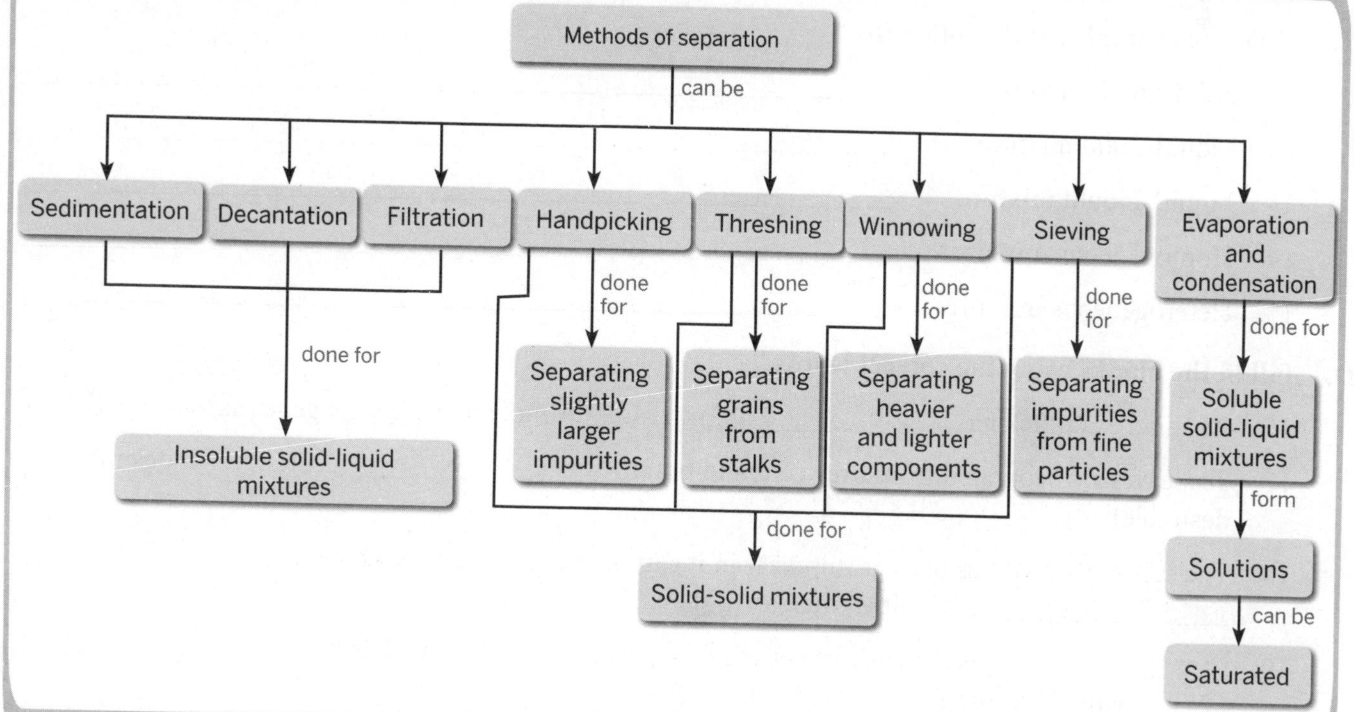

Worksheet 1

Name:

Grade: A B C D

A. Choose the correct option.

1. Which of the following is an appropriate method to separate flour from semolina?
   (a) Handpicking (b) Threshing (c) Winnowing (d) Sieving
2. Which of the following cannot be threshed to remove grains?
   (a) Oats (b) Pulses (c) Wheat (d) Barley
3. What is obtained on churning the milk?
   (a) Cream (b) Butter (c) Ghee (d) Oil
4. Which of the following is not an example of a pure substance?
   (a) Honey (b) Iron (c) Milk (d) Water
5. By which of the following methods can peanuts be separated from a mixture of pulses and rice?
   (a) Handpicking (b) Winnowing (c) Threshing (d) Sieving

B. State whether the following statements are true (T) or false (F).

6. Husk is the inner covering of some seeds and fruits.
7. In winnowing, difference in weight of the solids is the property used in separating a mixture.
8. Different components of a mixture have different chemical and physical properties.
9. Pebbles and stones can be separated from sand by winnowing.
10. The small pieces of husk that are separated from flour after milling is called bran.

C. Give one example for the following.

11. Solid-Solid mixture ____________________
12. Liquid-Solid mixture ____________________
13. Liquid-Liquid mixture ____________________
14. Homogeneous mixture ____________________
15. Heterogeneous mixture ____________________

D. Fill in the blanks using the correct words.

16. Air is an example of a ____________ mixture. (homogeneous/heterogeneous)
17. Tea leaves are separated from brewed tea to obtain ____________ substance. (undesirable/desirable)
18. The substance that has bran as impurity in it can be separated by the process of ____________. (handpicking/sieving)
19. ____________ is the easiest method of separation of wheat and its husk. (Winnowing/Threshing)
20. ____________ in water is considered as a heterogeneous mixture. (Salt/Oil)

Worksheet 2

Name:

Grade: (A) (B) (C) (D)

A. Answer the following questions in a sentence.

1. Name the method used to separate larger impurities from the mixture by handpicking. Give one example.

   ____

2. Give one example of sieving from daily life.

   ____

3. What is threshing?

   ____

4. What is the importance of the size of holes in a sieve?

   ____

5. Define a mixture.

   ____

B. Answer the following questions in brief.

6. Mention the different types of mixtures based on the state of components that are mixed together.

   ____

   ____

7. Write a short note on handpicking.

   ____

   ____

8. Explain the process of winnowing.

   ____

   ____

9. What is the criterion for the selection of a specific method of separation?

   ____

   ____

10. List the methods of separation of solid-solid mixtures.

   ____

   ____

C. Answer the following questions in detail.

11. Differentiate between a pure substance and a mixture.

____

____

____

12. Differentiate between homogeneous and heterogeneous mixtures.

____

____

____

13. What is the purpose of separation of mixtures?

____

____

____

14. You are given a bowl containing a mixture of grass and mint leaves. How will you separate them? Justify your answer.

____

____

____

15. Identify the type of separation in the image given below. Explain the method with an example.

____

____

____

Worksheet 3

Name: ____________

Grade: (A) (B) (C) (D)

A. Choose the correct option.

1. Which of these cannot be filtered from the water?
   (a) Dust (b) Sediments (c) Sand (d) Salts
2. Salt crystals undergo one of the following processes before being used.
   (a) Filtration (b) Purification (c) Decantation (d) Condensation
3. Which of these mixtures are separated by filtration?
   (a) Solid-Solid (b) Liquid-Liquid (c) Solid-Liquid (d) Liquid-Gas
4. Which of the following can dissolve in water?
   (a) Solids (b) Liquids (c) Gases (d) All of these
5. In a common salt solution, identify the solute.
   (a) NaOH (b) HCl (c) NaCl (d) H_2O

B. Fill in the blanks using the correct words.

6. Pulp from orange juice can be separated by ____________. (filtration/condensation)
7. Salt is obtained from seawater by ____________. (sedimentation/evaporation)
8. ____________ is the opposite process of evaporation. (condensation/decantation)
9. The layer of clear liquid above the sediment is called ____________. (supernatant/residue)
10. In soft drinks, carbon dioxide is one of the ____________. (solutes/solvents)

C. Give two examples for the following.

11. Solute ____________
12. Solvent ____________
13. Soultion ____________
14. Filtration of substances ____________
15. Condensation of substances ____________

D. State whether the following statements are true (T) or false (F).

16. Traditional methods of purification of water are more environmental-friendly.
17. We can dissolve more solute in a solvent at low temperature.
18. Sky becomes unclear after the first rain.
19. Water is considered as a universal solvent.
20. Sugar can be filtered from its solution.

Name: ____________

Grade: A B C D

Worksheet 4

A. Answer the following questions in a sentence.

1. What is sedimentation?

____

2. Give one example of filtration from daily life.

____

3. Define evaporation.

____

4. Define condensation.

____

5. What is a saturated solution?

____

B. Answer the following questions in brief.

6. List the different methods of separating solid-liquid mixtures.

____

____

7. What are the different solutes added during the preparation of tea?

____

____

8. How are gold particles found in the river separated from the sand?

____

____

9. What are the methods by which sand and water can be separated?

____

____

10. What happens when steam comes in contact with a cold metal plate?

____

____

C. Answer the following questions in detail.

11. Explain an activity to prepare a saturated solution.

12. How are husk or dirt particles from pulses separated before cooking? Explain.

13. Which sequence of processes is best suited for the separation of salt, sand and water?

14. What are the factors that can increase the solubility of a solute in a solvent?

15. Identify and explain the process of separation given below.

Worksheet 5

Name:

Grade: A B C D

A. Choose the correct option.

1. Which of the following is soluble in water?
   (a) Sand (b) Metal (c) Glass (d) Ink
2. In a soft drink, which one of these is not a solute?
   (a) Carbon dioxide (b) Sugar
   (c) Water (d) Salt
3. Which of the following methods is best suited for the separation of salt from water?
   (a) Filtration (b) Condensation (c) Evaporation (d) Sieving
4. Which of these dissolves most of the substances in it?
   (a) Water (b) Ink (c) Honey (d) Lemonade
5. Which among these methods is used to separate solid-solid mixture?
   (a) Filtration (b) Decantation (c) Winnowing (d) Condensation

B. Match the following.

6. Sand and water	(a) Evaporation and condensation
7. Flour and sawdust	(b) Filtration
8. Salt and water	(c) Handpicking
9. Candies and buttons	(d) Sedimentation and decantation
10. Fruit pulp and juice	(e) Sieving

C. Give reasons for the following.

11. Sugar and rice cannot be separated by handpicking.

    ____

12. Salt obtained directly from the salt pans cannot be used for various purposes.

    ____

13. The pores of the filter should be smaller than the size of the solid particles to be separated.

    ____

14. Salt dissolves in water to give a clear solution.

    ____

15. Winnowing cannot be done in the absence of air.

    ____

D. Give one example for the following.

16. Saturated solution ______________________
17. Unsaturated solution ______________________
18. Handpicking ______________________
19. Sieving ______________________
20. Evaporation ______________________

E. Answer the following questions in detail.

21. How is common salt obtained from seawater?

____

____

____

22. Explain an activity to separate a mixture of sand and water using sedimentation and decantation methods.

____

____

____

23. Explain an activity to separate a mixture of sand and water using filtration method.

____

____

____

24. Explain an activity to show that the amount of solute dissolved in the same amount of water is different for different substances.

____

____

____

25. Label the given diagram. Identify the separation technique and explain the same.

____

____

____

4 Fibre to Fabric

Recap of the Chapter

- Evolution in clothing began with the stone age man.
- Agriculture and industrial revolution in the 19th century influenced the weaving industry worldwide.
- Though the clothing industry has grown tremendously in 20th and 21st century, unstitched fabric is still worn by many people.
- Fabric is a piece of cloth made from fibre yarn.
- Fabric is classified into natural and synthetic or man-made based on the source of the fibre used.
- Natural fibres include plant fibres and animal fibres.
- Fabric from yarn is produced using two processes called weaving and knitting.

Concept Map

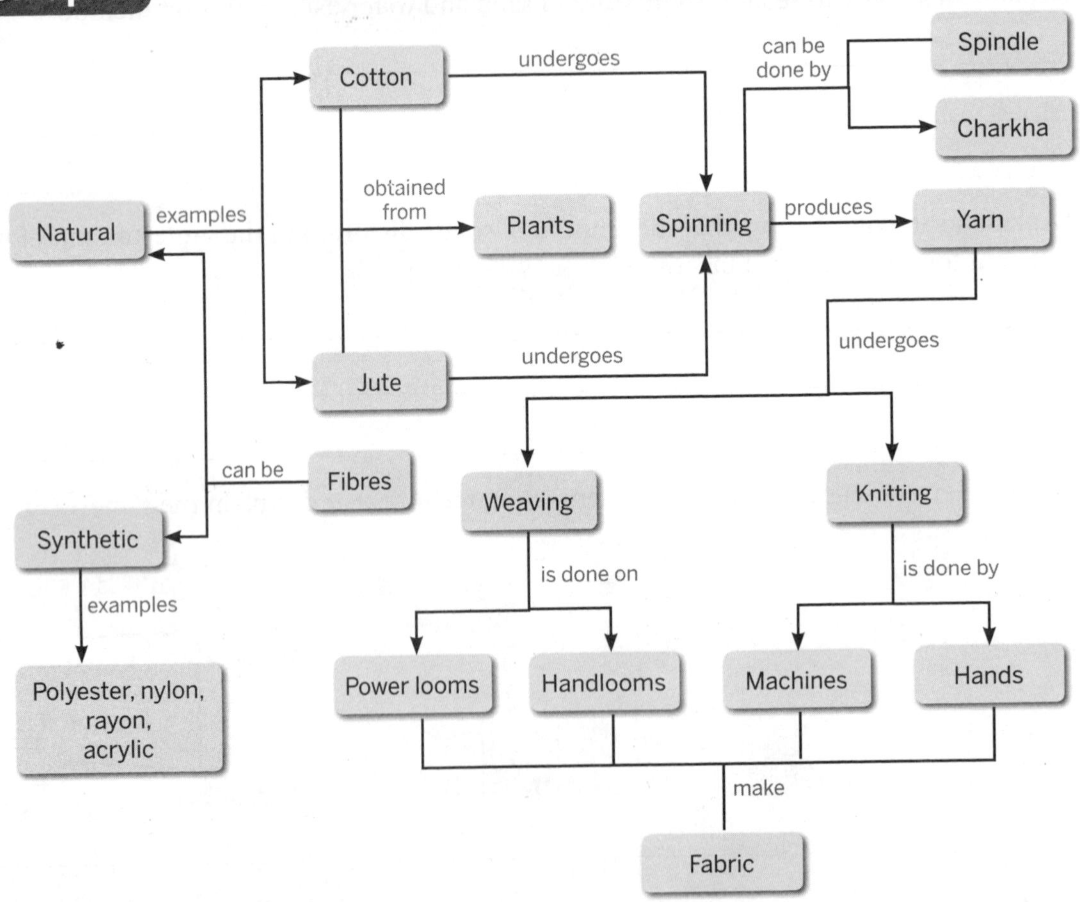

Worksheet 1

Name: ____________

Grade: A B C D

A. Choose the correct option.

1. Which of the following is true for a silk sari?
   (a) Unstitched fabric (b) Stitched fabric
   (c) Man-made fibre (d) Blended fabric
2. Which of the following fibres is obtained from the stem of flax plant?
   (a) Cotton (b) Flax (c) Hemp (d) Coir
3. Which of the following is obtained from animals?
   (a) Jute (b) Sisal (c) Silk cotton (d) Wool
4. Which of the following is an example of synthetic fibre?
   (a) Acrylic (b) Rayon (c) Polyester (d) All of these
5. Which of the following is a hair-like structure?
   (a) Fabric (b) Yarn (c) Fibre (d) None of these

B. Fill in the blanks using the correct words.

6. The fabric that absorbs more water is ____________. (cotton/wool)
7. Jute is a ____________ fibre. (natural/artificial)
8. Earlier people wore ____________ clothes. (stitched/unstitched)
9. Around 40,000 years ago, needles were made from ____________. (bones/steel)
10. Gunny bags and ropes are made from ____________. (jute/wool)

C. Name the following.

11. The first country known to have grown cotton ____________
12. The inventor of weaving machine ____________
13. Strand used to make yarn ____________
14. Material used to make cloth ____________
15. Fibre obtained from sheep ____________

D. State whether the following statements are true (T) or false (F).

16. Clothes protect us from harsh weather conditions.
17. People wear same type of clothes during different seasons.
18. Synthetic fibres are also called man-made fibres.
19. Animal fibres are not natural fibres.
20. Wet clothes do not dry fast during summer.

Worksheet 2

Name: ____________

Grade: A B C D

A. Answer the following questions in a sentence.

1. How did stone age man use to cover his body?

   ____

2. Which method is used to produce coir yarn?

   ____

3. What do you get if a yarn is split along its length?

   ____

4. During which century did western clothing gain popularity in India?

   ____

5. Which fibre is obtained from the cocoons of silk moth?

   ____

B. Answer the following questions in brief.

6. How many types of looms are there? Name them.

   ____

   ____

7. What is the use of wearing clothes?

   ____

   ____

8. Why did early man use tree bark, leaves and animal skin to cover his body?

   ____

   ____

9. Describe the process of making synthetic fibres with an example.

   ____

   ____

10. What is microfibre?

    ____

    ____

C. **Answer the following questions in detail.**

11. Describe the clothing used by man from past to present.

12. Explain the process of making a fabric from fibre.

13. Explain the process of making cotton yarn from cotton.

14. How did agricultural and industrial revolution affect the weaving industry?

15. What are plant fibres? Describe the two types of cotton fibres.

Worksheet 3

Name: ____________

Grade: A B C D

A. Choose the correct option.

1. Which of the following natural fibres is used in the textile industry?
   (a) Cotton (b) Rayon (c) Nylon (d) Polyester
2. Any fabric that is put to a specific use is called
   (a) Cloth (b) Yarn (c) Fibre (d) Weave
3. A natural fibre that is inexpensive is
   (a) Wool (b) Silk (c) Cotton (d) Jute
4. Which of the following was popularized by Gandhi?
   (a) Spinning jenny (b) Charkha (c) Takli (d) Handloom
5. Name the machine used for weaving of the fabric.
   (a) Cotton gin (b) Hand spindle (c) Loom (d) Spinning wheel

B. State whether the following statements are true (T) or false (F).

6. Two sets of yarn are used during weaving.
7. Natural fibres burn and produce bright red flame.
8. A bale of cotton contains bundle of fibres.
9. Mature cotton fruit are yellow in colour.
10. Jute stem are harvested during the flowering stage.

C. Complete the pattern.

11. Jute fibre : Outer part of stem :: Cotton fibre : ____________
12. Bale : Cotton yarn :: Rotten jute stem : ____________
13. Cotton : Ginning :: ____________ : Retting
14. Power loom : ____________ :: Handloom : Human muscle power
15. Handloom textile : Handloom machine :: Woollen fabric : ____________

D. Fill in the blanks using the correct words.

16. Jute plants grow well in ____________ soil. (alluvial/sandy)
17. The stems of ____________ plants are allowed to rot to extract fibre. (jute/cotton)
18. In large scale production of yarn, ____________ is used. (spinning machine/takli)
19. Power loom is used in the process of ____________ . (knitting/weaving)
20. Through the process of ginning, ____________ fibres are separated from seeds. (cotton/silk)

Worksheet 4

Name:

Grade:

A B C D

A. Answer the following questions in a sentence.

1. Define knitting and name the device used for manual knitting of the yarn.

2. Name two devices used to spin yarn.

3. What is ginning?

4. Define weaving.

5. Give one difference between knitting and weaving.

B. Answer the following questions in brief.

6. Can different fabrics be knitted? If yes, give examples.

7. Describe the use of takli and charkha in the process of hand spinning.

8. Identify the hand tool in the given picture and describe its use.

9. What is meant by braiding a yarn?

10. Describe the process of weaving and its types.

C. Answer the following questions in detail.

11. Explain the two methods of knitting.

____

____

____

12. Name two machines used in the weaving industry and write any two differences between them.

____

____

____

13. Why is cotton grown mainly in Maharashtra, Gujarat and Madhya Pradesh?

____

____

____

14. Explain the process of making cotton fabric from cotton bolls.

____

____

____

15. Describe the similarities between the given images.

A

B

Image A	Image B

Worksheet 5

Name: ____________

Grade: A B C D

A. Choose the correct option.

1. Kapok tree gives us one of the following fibres.
   (a) Silk cotton (b) Nylon (c) Rayon (d) Coir
2. An electric spinning machine is
   (a) Spinning Jenny (b) Takli
   (c) Charkha (d) Handloom
3. Which of the following fibres catches fire easily?
   (a) Jute (b) Silk (c) Yak wool (d) Sheep wool
4. Which of the following is grown in Madhya Pradesh?
   (a) Sugarcane (b) Cotton (c) Coconut (d) Rice
5. Which of the following fibres is used to make inexpensive ropes?
   (c) Polyester (b) Jute (c) Silk (d) Rayon

B. Match the following.

6. Shrub grown on black soil	(a) Ginning
7. Separating cotton seeds from fibre	(b) Cotton
8. Handpicking jute fibre	(c) Power loom
9. Making yarn from fibre	(d) Retting
10. Electricity-run loom	(e) Spinning

C. Give reasons for the following.

11. There is a change in the trends of clothing during different times.

____

12. Hemp and jute have a similarity.

____

13. When a part of the woollen shawl is burnt, it smells like burnt hair.

____

14. Polycot fabric is also known as blended fabric.

____

15. Coir is a natural fibre.

____

D. Give one example for the following.

16. Fibre obtained from the stem of hemp plant __________
17. Fibre obtained from the stem of flax plant __________
18. Fibre obtained from the cocoons of silk moth __________
19. Fibre obtained from the hair of sheep __________
20. Fibre obtained from the fruit of kapok trees __________

E. Answer the following questions in detail.

21. Clothes are as important as food and shelter for human beings.

22. Identify and write the names of the following machines. Write one use of each machine in weaving industry.

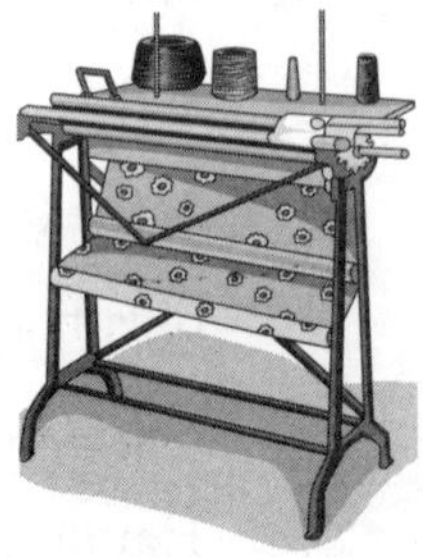

23. Explain the process of knitting with an example.

24. Differentiate between weaving and knitting.

25. Differentiate among yarn, fibre and fabric.

5 Sorting Materials

Recap of the Chapter

- Sorting or arranging materials help us to find things easily when we need them.
- Things are grouped on the basis of similarities or differences in their properties.
- Materials can be classified based on their state into three categories: solids, liquids and gases.
- Based on the texture, materials are classified into rough and soft materials.
- Solubility is the property of a substance to dissolve in a given liquid.
- Materials that are denser than water will sink in water, whereas materials that are less dense than water will float on water.
- Miscible liquids mix with water completely, whereas immiscible liquids do not mix water.
- Depending on the amount of light that can pass through materials, they can be classified into three categories: transparent, translucent and opaque.

Concept Map

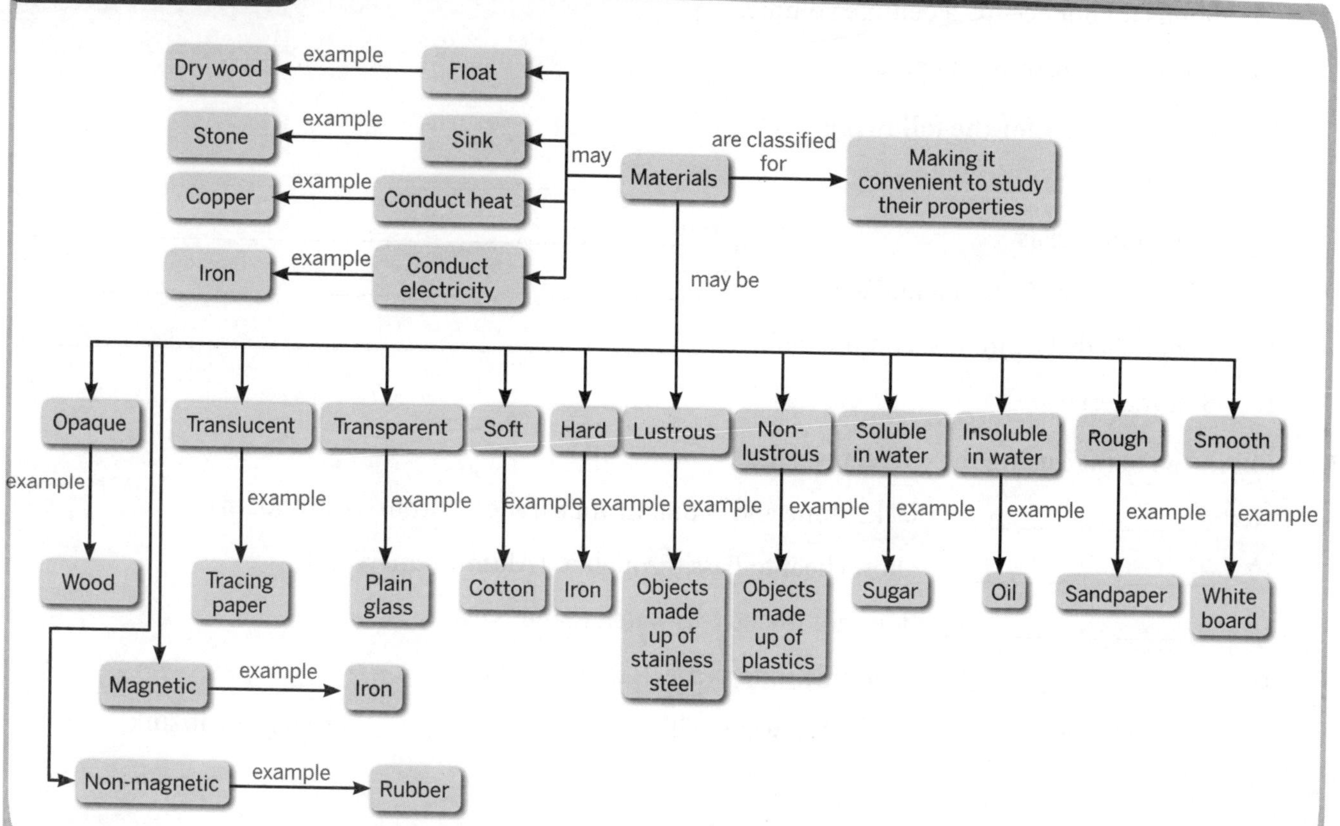

Worksheet 1

Name:

Grade: A B C D

A. Choose the correct option.

1. Which of the following substances is dull in appearance?
   (a) Bronze (b) Silver (c) Cardboard (d) Gold
2. Which of the following materials can be compressed easily?
   (a) Sandpaper (b) Wool (c) Wood (d) Body scrubber
3. Identify the material that has a rough texture.
   (a) Sponge (b) Sandpaper (c) Foam (d) Wool
4. Identify the substance that has a definite shape and occupies a definite volume.
   (a) Eraser (b) Sugar solution (c) Fruit juice (d) Tea
5. Which of the following is highly compressible?
   (a) Sugar crystals (b) Plastic box (c) Oxygen (d) Fruit juice

B. State whether the following statements are true (T) or false (F).

6. Lustrous materials can lose their shine when exposed to moisture for a long duration.
7. Cotton is difficult to compress.
8. Gases are highly compressible.
9. Solids do not occupy a definite volume.
10. Bronze is a lustrous material.

C. Give two examples for the following.

11. Lustrous materials ______________________
12. Rough materials ______________________
13. Materials that exist in liquid state ______________________
14. Materials that cannot be compressed ______________________
15. Soft materials ______________________

D. Fill in the blanks using the correct words.

16. ______________ lacks shine and is dull in appearance. (Bronze/Cardboard)
17. ______________ allows heat to pass through it. (Aluminium/Wood)
18. ______________ does not dissolve in water. (Sand/Salt)
19. ______________ have a definite shape. (Gases/Solids)
20. The material that is less dense than water will ______________ on water. (float/sink)

Worksheet 2

Name:

Grade:

Ⓐ Ⓑ Ⓒ Ⓓ

A. Answer the following questions in a sentence.

1. Name three substances that do not change their volume and shape when transferred from one container to another.

   ____

2. Define the property of a material to shine.

   ____

3. Name three materials that are hard and are difficult to compress.

   ____

4. Name two substances that do not have a definite shape and volume.

   ____

5. Define texture.

   ____

B. Answer the following questions in brief.

6. Define classification.

   ____

   ____

7. Mention the importance of classification of materials.

   ____

   ____

8. What will happen if iron (lustrous material) is rubbed with sandpaper?

   ____

   ____

9. What kind of objects float on the surface of water? Give two examples.

   ____

   ____

10. Identify the common property (in terms of its texture) that the following materials have.
    (a) Wood (b) Rubber (c) Plastic

    ____

    ____

C. **Answer the following questions in detail.**

11. What will happen to the volume of 50 ml tea contained in a kettle if it is transferred to a cup?

12. Identify the state of the material based on the arrangement of the particles shown in the image below. Write the properties of this state.

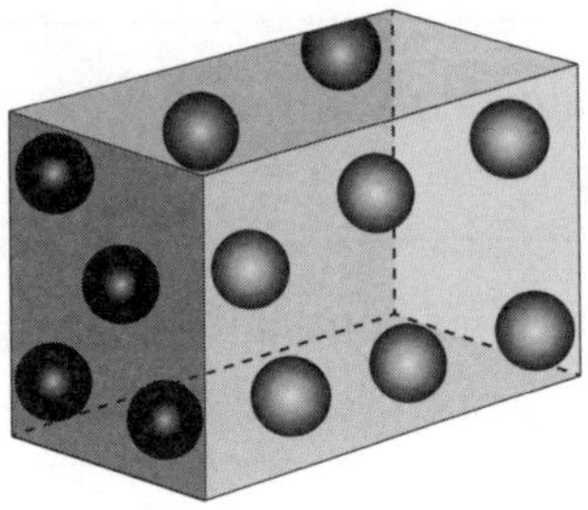

13. Distinguish between lustrous and non-lustrous materials with an example.

14. Madhu left a silver article and an article made of rubber for a long duration in the moisture. Would there be any change in the articles? If so, why?

15. Draw the arrangement of particles of the following materials.
(a) Wooden block (b) Coloured liquid (c) An inflated balloon

Worksheet 3

Name: ______

Grade: A B C D

A. Choose the correct option.

1. Which of the following substances is insoluble in water?
   (a) Salt (b) Sugar crystals (c) Sand (d) Sugar cubes
2. Which of the following is immiscible in water?
   (a) Alcohol (b) Glycerin (c) Oil (d) Milk
3. Identify the translucent material from the following.
   (a) Cardboard (b) Wood (c) Frosted glass (d) Clear glass
4. Which of the following dissolves in water completely?
   (a) Diesel (b) Oxygen (c) Petrol (d) Coconut oil
5. Which of the following is denser than water?
   (a) Feather (b) Dried leaf (c) Stone (d) Wood

B. State whether the following statements are True (T) or False (F).

6. Light does not pass through an opaque object.
7. Big objects sink in water and small objects float on water.
8. Mud sinks in water.
9. Objects made of aluminium do not allow heat to pass through them.
10. Gases do not dissolve in liquids.

C. Give two examples for the following.

11. Opaque materials ______
12. Insulators ______
13. Non-magnetic materials ______
14. Translucent materials ______
15. Miscible liquids: ______

D. Match the following.

16. Translucent	(a) Plastic toy
17. Opaque	(b) Vinegar
18. Magnetic	(c) Stone
19. Non-magnetic	(d) Iron key-chain
20. Miscible liquid	(e) Fog

Worksheet 4

Name:

Grade: A B C D

A. Answer the following questions in a sentence.

1. Name two substances that are completely soluble in water.

2. What are miscible liquids?

3. What type of materials get attracted towards a magnet? Give two examples.

4. Define an insulator.

5. Is mirror a transparent object? Why or why not?

B. Answer the following questions in brief.

6. Differentiate between transparent and opaque objects.

7. Distinguish between miscible and immiscible liquids.

8. What is solubility? Explain by taking appropriate examples.

9. Mention the difference between solubility of the following substances in water.
   (a) Vinegar (b) Kerosene

10. Distinguish between conductors and insulators of electricity.

C. Answer the following questions in detail.

11. What is density? Explain with an example.

12. What is the property of the material shown in the given image? Write a note on it.

13. Which property of materials is shown in the given image? What kind of substances show this property? Give three examples of substances showing this property.

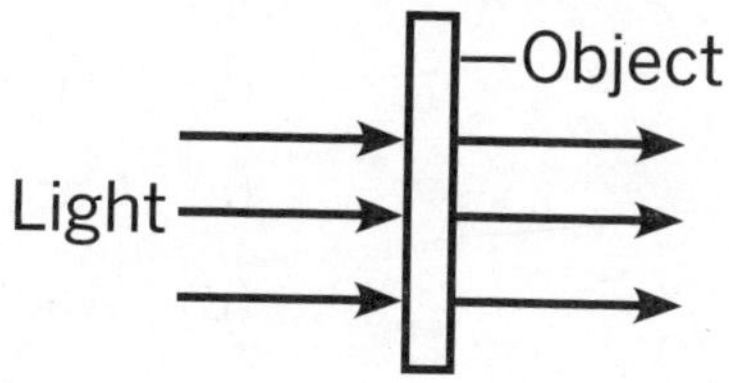

14. How can materials be classified based on their conductivity?

15. Mention the category to which the following materials belong and also write their property.
(a) Clear glass (b) Wood (c) Butter paper

Worksheet 5

Name:

Grade: A B C D

A. Choose the correct option.

1. Knife and needle belong to the same category because they are
   (a) magnetic (b) non-lustrous
   (c) miscible (d) insulators
2. Identify the property that does not describe the physical property of the materials.
   (a) Smoothness (b) Rusting (c) Hardness (d) Lustre
3. In water, an empty plastic bottle
   (a) Sinks (b) Floats (c) Corrodes (d) Reacts
4. Butter paper is an example of
   (a) opaque material (b) transparent material
   (c) translucent material (d) magnetic material
5. Which of the following is denser than water?
   (a) Cotton (b) Aluminium sheet (c) Empty plastic bottle (d) A piece of wood

B. Match the following.

6. Soft material	(a) Stone
7. Hard material	(b) Plastic
8. Shiny material	(c) Copper
9. Conductor	(d) Wool
10. Insulator	(e) Gold

C. Give reasons for the following.

11. The cooking utensils have plastic covering over them.

____

12. Dried leaves float on water but the stone sinks in water.

____

13. An object made of rubber does not get attracted towards a magnet.

____

14. We can see clearly through a plain glass but not through a coloured glass.

____

15. We can see a separate layer of coconut oil floating in a beaker containing water.

____

D. **Give one example for the following.**

16. Non-lustrous material ____________________
17. Smooth material ____________________
18. Immiscible liquid ____________________
19. Magnetic material ____________________
20. Transparent material ____________________

E. **Answer the following questions in detail.**

21. Identify the property of the material shown in the given picture and write a note on it.

____

____

____

22. Write the properties of the materials shown in the images given below.

(a)

(b)

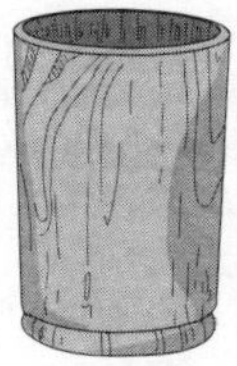

(c)

____

____

____

23. Can liquids be dissolved in water? If so, explain it with an example.

____

____

____

24. Explain any three properties of materials with examples.

____

____

____

25. Write the properties of three states of matter.

____

____

____

6 Changes Around Us

Recap of the Chapter

- Reversible changes are the changes in which we can get back the original substance by changing the conditions.
- Irreversible changes are the changes in which we cannot get back the original substance.
- Physical changes are the changes in which no new substances are formed.
- Chemical changes are the changes in which new substances are formed.
- The increase in the size of a substance on heating is called expansion.
- The decrease in the size of a substance on cooling is called contraction.

Concept Map

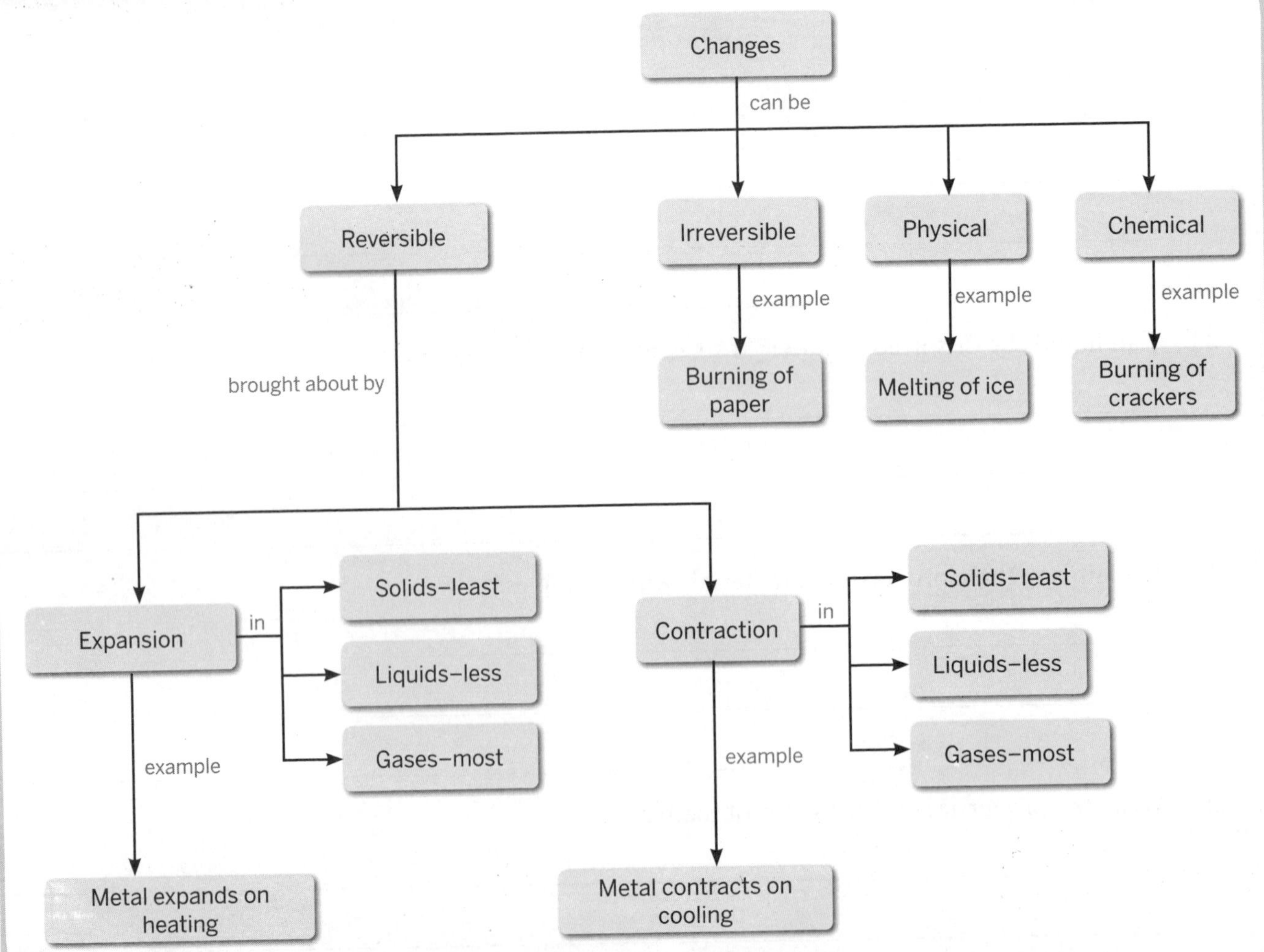

Worksheet 1

Name: ____________

Grade: A B C D

A. Choose the correct option.

1. Which of the following is a reversible physical change?
   (a) Melting of ice (b) Burning of a paper (c) Growth of a plant (d) Tearing of a paper
2. Which among the following is not a physical change?
   (a) Burning (b) Melting (c) Condensation (d) Evaporation
3. Identify the reversible change from the following.
   (a) Curdling of milk (b) Cooking of rice
   (c) Growth of a plant (d) Heating a metal rod
4. Melting of glaciers is an example of
   (a) irreversible change (b) physical change
   (c) chemical change (d) expansion
5. Which among the following is true about reversible changes?
   (a) Change is temporary (b) Change is permanent
   (c) Both (a) and (b) (d) Neither (a) nor (b)

B. State whether the following statements are true (T) or false (F).

6. All changes in nature are physical in nature.
7. Expansion of metals on heating is a reversible change.
8. We can get back the salt that has been dissolved in water.
9. In reversible changes no new substance is formed.
10. The properties of the substance change in reversible changes.

C. Give one example for the following.

11. Reversible change ____________
12. Physical change ____________
13. Physical change but not reversible ____________
14. A reversible physical change ____________
15. A change in the shape of the substance ____________

D. Fill in the blanks using the correct words.

16. ____________ are the changes in which we can get back the original substance by changing the conditions. (Reversible changes/ Irreversible changes)
17. The process of freezing is ____________ change. (a reversible/an irreversible)
18. Rusting of iron is ____________ change. (a reversible/an irreversible)
19. On heating, mercury ____________. (contracts/ expands)
20. Substances can decrease in size on ____________. (heating/cooling)

Worksheet 2

Name:

Grade: A B C D

A. Answer the following questions in a sentence.

1. How can we get solid ice back from water?

2. Why is dissolving sugar in water a physical change?

3. Why is burning of paper not a reversible change?

4. Stretching of a rubber band is not an irreversible change. Why?

5. What are the characteristics of a substance that undergoes a change?

B. Answer the following questions in brief.

6. How can we judge that a change has occurred?

7. How can we differentiate between the following changes?
   (a) Milk changes into curd (b) Ice melts to form water

8. Write a short note on reversible changes.

9. What is meant by physical changes?

10. In physical changes, chemical properties of the substance remain the same. Justify with appropriate examples.

C. **Answer the following questions in detail.**

11. How can we say that water cycle involves reversible changes?

12. Observe the pictures given below. Identify the picture/s that represent/s physical change. Justify your answer.

(a) (b) (c)

13. Explain an activity that shows physical change.

14. There is a change in the candle's height when it is burning. Can this change be reversed? Justify your answer.

15. All reversible changes are physical changes, but all physical changes need not be reversible. Justify.

Worksheet 3

Name: ______

Grade: A B C D

A. Choose the correct option.

1. Ripening of fruits is __________ change.
   (a) a reversible (b) an irreversible (c) a physical (d) None of these
2. Which among the following is an example of a chemical change?
   (a) Stretching of rubber (b) Freezing (c) Boiling of water (d) Digestion of food
3. Which of the following processes can lead to irreversible change?
   (a) Growth (b) Cooling (c) Dissolving (d) Heating
4. Which among the following is not an irreversible change?
   (a) Burning of wood (b) Rotting of wood (c) Evaporation (d) Cooking rice
5. Which of the following is true about irreversible changes?
   (a) Change is permanent (b) Change is temporary
   (c) Both (a) and (b) (d) Neither (a) nor (b)

B. State whether the following statements are true (T) or false (F).

6. Curdling of milk is an example of a reversible change.
7. An irreversible change produces new substances.
8. Rusting of iron is a slow process and it is an irreversible change.
9. The process of sublimation involves chemical change.
10. The extent to which substances expand differs for solids, liquids and gases.

C. Give one example for the following.

11. Irreversible change ______________
12. Chemical change ______________
13. Irreversible physical change ______________
14. Expansion ______________
15. Contraction ______________

D. Fill in the blanks using the correct words.

16. Heating of iodine crystals is a __________ change. (physical/chemical)
17. Burning of wood is __________ change. (a reversible/an irreversible)
18. Digestion of food involves many __________ changes inside our body. (physical/chemical)
19. In a __________ changes, both physical and chemical properties of the original substance change. (chemical/physical)
20. Dissolving sugar in water is a __________ change as there is no formation of any new substance. (chemical/physical)

Worksheet 4

Name: ____________

Grade: A B C D

A. Answer the following questions in a sentence.

1. Why is burning of paper not a reversible change?

____

2. Why chemical change is a permanent change?

____

3. Why is the process of heating not an irreversible change?

____

4. The growth of an organism is an irreversible change. Why?

____

5. When milk boils, it spills out of the container if the burner is not turned off. Why?

____

B. Answer the following questions in brief.

6. What are irreversible changes?

____

____

7. How do we know that a chemical change has taken place?

____

____

8. Write a note on chemical changes.

____

____

9. What is meant by contraction?

____

____

10. What do you mean by expansion?

____

____

C. **Answer the following questions in detail.**

11. Which among the following figures represents irreversible change? Justify your answer and give two examples of chemical change that are not reversible.

(a)

(b)

(c)

12. Differentiate between expansion and contraction with suitable examples.

13. Explain how an iron blade, the head of an axe, is fixed onto its wooden handle to be used as a tool to cut the wood.

14. A steel vessel is stuck with another one. How will you separate the vessels and explain the science behind it?

15. Write any three real-life applications of expansion and contraction.

...ksheet 5

Name:

Grade: A B C D

A. Choose the correct option.

1. Pick the odd one out.
   (a) Burning of petrol
   (b) Lighting of a sparkler
   (c) Mixing of cement with water
   (d) Boiling of water
2. Reversible change: Folding of paper :: Irreversible change: ______________
   (a) Salt mixing with water
   (b) Melting of chocolate
   (c) Boiling of water
   (d) Tearing of paper
3. During heating, particles of the substance will
   (a) move away (b) come closer (c) remain fixed (d) None of these
4. During contraction, the particles of the substance will
   (a) move away (b) come closer (c) remain fixed (d) None of these
5. Identify the correct order related to expansion of substances.
   (a) Solid > Gas > Liquid
   (b) Gas > Solid > Liquid
   (c) Gas > Liquid > Solid
   (d) Liquid > Gas > Solid

B. Match the following.

6. Growth of animals	(a) Contraction
7. Heating of ice	(b) Expansion
8. On cooling, particles are tightly packed	(c) Irreversible change
9. On heating, particles are loosely arranged	(d) Purple coloured vapour
10. Iodine crystals on heating	(e) Physical change

C. Give reasons for the following.

11. Blacksmiths make iron tools by heating the iron to a certain temperature.

____

12. Metal rims on the wheels of a wooden cart can be fitted using the principle of both contraction and expansion.

____

13. Railway tracks are laid with gaps in between them.

____

14. Tightly jammed metal lids of food jars can be opened easily by pouring hot water over the lid.

____

15. When milk boils, it spills out of the container if the burner is not turned off.

____

D. State whether the following statements are true (T) or false (F).

16. In certain reversible processes, the actual material cannot be obtained back.
17. Heating leads to reversible change and burning leads to irreversible change.
18. Process of change of a solid into liquid is called melting.
19. Melting of wax is an irreversible change whereas burning of wax is a reversible change.
20. Combination of two substances to form a new substance is a reversible process.

E. Answer the following questions in detail.

21. Explain the differences between physical and chemical changes with suitable examples.

22. The figure given below is the experimental set-up to identify the type of change that takes place when iodine crystals are heated. Explain the procedure involved in it.

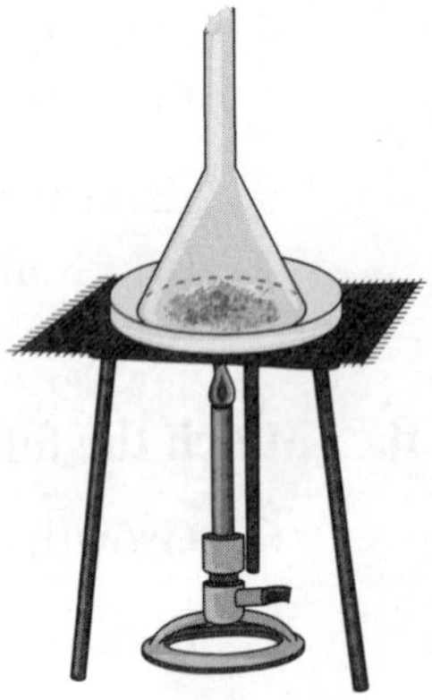

23. Differentiate between reversible and irreversible changes with examples.

24. Explain the procedure of an experiment that shows air contracts on cooling.

25. The extent to which substances expand or contract differs for solids, liquids and gases. Justify.

7 Things Around Us

Recap of the Chapter

- All objects in nature can be divided into living and non-living.
- Living things have some characteristics that make them different from non-living things.
- Living things need food to survive, to grow in height and size.
- All living things can move from one place to another except for the plants, as they are rooted to the soil.
- Living things respire, excrete and reproduce.
- Living things respond to stimuli.
- Living things show different levels of structural organization and has a life span.
- A place where plants and animals live naturally is called a habitat.
- All living things form the biotic components and non-living things form the abiotic components of a habitat.

Concept Map

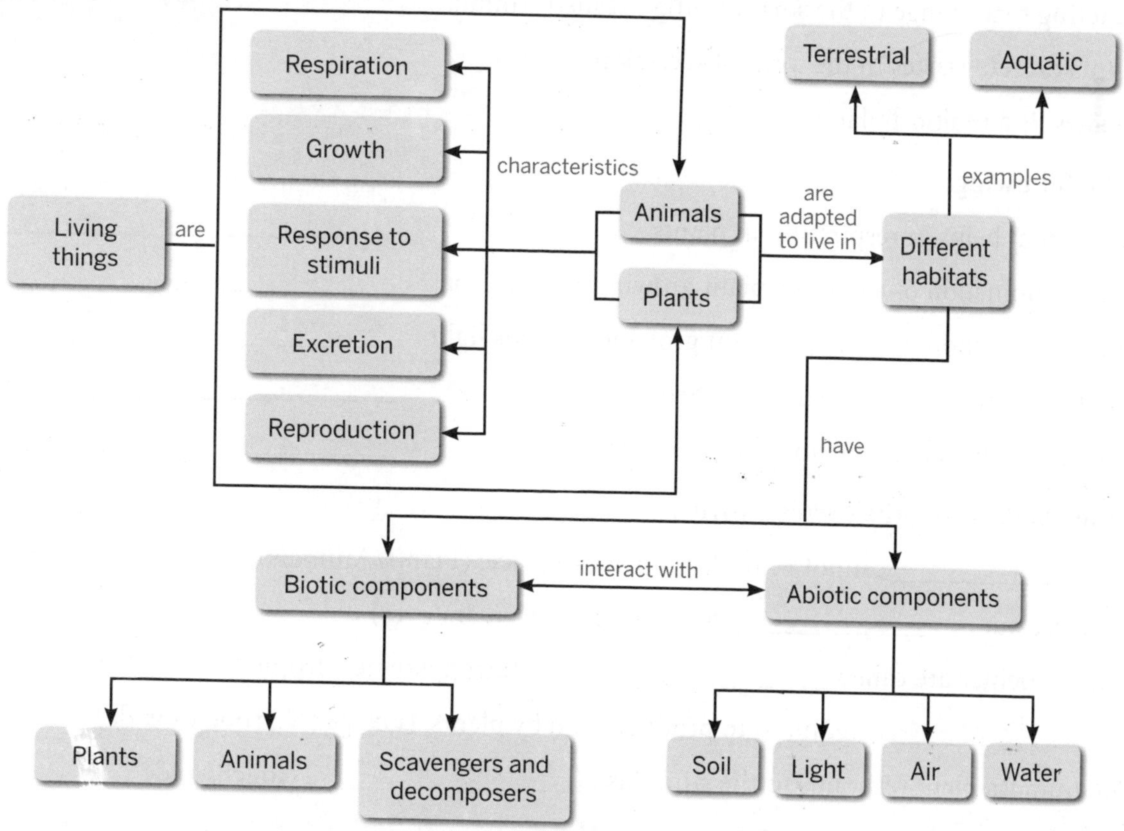

Worksheet 1

Name: ____________

Grade: (A) (B) (C) (D)

A. Choose the correct option.

1. Which of the following is a unicellular organism?
   (a) *Euglena* (b) Plants (c) Animals (d) Human beings
2. Which of these animals gives birth to young ones?
   (a) Snake (b) Hen (c) Cow (d) Duck
3. Identify the organ that is a part of circulatory system?
   (a) Lungs (b) Stomach (c) Liver (d) Heart
4. Which of the following is a source of stimuli?
   (a) Touching a needle (b) Drinking water
   (c) Eating food (d) Sitting on chair
5. Which is the smallest part in a human body?
   (a) Tissue (b) Cell (c) Stomach (d) Skin

B. State whether the following statements are true (T) or false (F).

6. All living things need food for their survival and growth.
7. Living things breathe in carbon dioxide and breathe out oxygen during respiration.
8. Reacting to a change in the surrounding is called stimuli.
9. Plants excrete wastes in the form of secretions.
10. *Amoeba* is a multicellular organism.

C. Name the following.

11. Organ that helps in respiration in plants ____________
12. The phenomenon of roots of a plant growing towards soil ____________
13. The phenomenon of stem of a plant growing towards light ____________
14. Plant that curls its leaves when touched ____________
15. Organisms that are made up of several cells ____________

D. Fill in the blanks using the correct words.

16. ____________ cannot move from place to place. (Plants/Animals)
17. We close our ____________ to the flashlights. (ears/eyes)
18. Human beings are called ____________. (autotrophs/heterotrophs)
19. ____________ is required to produce food by plants. (Oxygen/Carbon dioxide)
20. The mimosa plant responds to the stimulus of ____________. (touch/cold)

Worksheet 2

Name:

Grade:

A B C D

A. Answer the following questions in a sentence.

1. What is geotropism?

2. Define respiration.

3. What is excretion?

4. Give example of a response and stimulus.

5. Define and name two unicellular organisms.

B. Answer the following questions in brief.

6. Name the levels of organization in the human body in the order from smallest to largest.

7. Define life cycle.

8. How do plants breathe?

9. List any four stimuli.

10. Write the different stages in the life cycle of a hen in sequence.

C. Answer the following questions in detail.

11. Differentiate between autotrophs and heterotrophs.

12. Write a short note on stimuli and response.

13. What are the different ways of excretion in animals?

14. Explain the different types of movements in plants.

15. Observe the given image and comment on it.

Worksheet 3

Name: ____________

Grade: A B C D

A. Choose the correct option.

1. Which of these is a biotic component of a habitat?
   (a) Sand (b) Paper (c) Butterfly (d) Water
2. Identify the decomposer.
   (a) Maggots (b) Fungi (c) Condor (d) Jackal
3. Which of the following is not associated with the process of photosynthesis?
   (a) Rain (b) Oxygen (c) Chlorophyll (d) Carbon monoxide
4. Which of these abiotic components forms the topmost layer of the Earth's crust?
   (a) Soil (b) Water (c) Air (d) Light
5. Which of these is the main source of energy to the Earth?
   (a) Sunlight (b) Rainwater (c) Wind (d) Soil

B. Fill in the blanks using the correct words.

6. The green coloured pigment found in leaves is ____________. (chlorophyll/chloroplast)
7. ____________ feed on dead animals.(Omnivores/Scavengers)
8. ____________ carries oxygen and carbon dioxide in the dissolved form. (Water/Soil)
9. ____________ increases the fertility of the soil. (Water/Humus)
10. Water bodies such as lakes, rivers, seas and oceans form ____________ habitat. (terrestrial/aquatic)

C. Name the following.

11. Organisms that make their own food ____________
12. Animals that feed only on plant sources ____________
13. Animals that feed only on the flesh of other animals ____________
14. Animals that feed on both plants and animals ____________
15. Animals that feed on the remains of dead animals ____________

D. State the following statements are true (T) or false (F).

16. Plants require abiotic components to survive.
17. Decomposers are not important in a habitat.
18. Sunlight is the main source of energy.
19. Climatic changes do not effect the formation of soil.
20. A hibiscus plant can survive in an aquatic habitat.

Worksheet 4

Name:

Grade: A B C D

A. Answer the following questions in a sentence.

1. What is a habitat?

____

2. What are the two components of a habitat?

____

3. Define photosynthesis.

____

4. What are the components in air that are essential for the survival of living things?

____

5. How is soil formed?

____

B. Answer the following questions in brief.

6. How is humus formed?

____

____

7. List the abiotic components of a wetland habitat.

____

____

8. What are the conditions that play a major role as abiotic components that influence the biotic components?

____

____

9. Explain the role of water in the survival of biotic components.

____

____

10. Explain the role of light in a habitat.

____

____

C. Answer the following questions in detail.

11. Write a note on different groups of animals in a habitat.

12. Differentiate between scavengers and decomposers.

13. How does soil help plants to survive?

14. Explain the role of stomata and chlorophyll in the process of photosynthesis in plants.

15. Label the given diagram.

Worksheet 5

Name:

Grade: A B C D

A. Choose the correct option.

1. Which is the largest organ of excretion in the human beings?
   (a) Kidney (b) Nose (c) Skin (d) Large intestine
2. Which among the following is an autotroph?
   (a) Sheep (b) Goat (c) Grass (d) Lamb
3. Which among the following hatches from an egg?
   (a) Dog (b) Cat (c) Ostrich (d) Squirrel
4. Which of these is a response?
   (a) Cry (b) Smell (c) Heat (d) Cold
5. *Amoeba* : Unicellular :: Swan:: ____________________
   (a) Autotroph (b) Phototroph (c) Heterotroph (d) Multicellular

B. State whether the following statements are true (T) or false (F).

6. Growth begins from a single cell.
7. Stomata are present mostly on the upper side of leaves.
8. Every living thing responds in the same way to all the stimuli.
9. Microorganisms constitute the biotic component of a habitat.
10. Sunrays help only in the regulation of body temperature in animals.

C. Give reasons for the following.

11. Cutting of leaves may kill the plant.

    ____

12. Reproduction is seen only among the biotic components.

    ____

13. Plants and animals mostly share a common habitat for their existence.

    ____

14. Roots of a plant grow deep down into the soil.

    ____

15. Heterotrophs are dependent on autotrophs for their survival.

    ____

D. Give two examples for the following.

16. Stimuli ____________________________
17. Organ systems ____________________________

18. Decomposers ______________________________

19. Habitat ______________________________

20. Abiotic components ______________________________

E. Answer the following questions in detail.

21. List the characteristics of living things.

____

____

____

22. How do living things reproduce?

____

____

____

23. Observe the given image and comment on it.

____

____

____

24. Write an activity to show the growth of plants from seeds.

____

____

____

25. Label the image given below to show how the biotic and abiotic components interact in nature.

8 Plants

Recap of the Chapter

- Root system of a plant constitutes the root, its branches, root hair and root cap.
- Plants have two types of root systems: tap root system and fibrous root system.
- Shoot system of a plant includes stem, leaf, flower and fruit, etc.
- Stems of plants get modified for storage, reproduction, protection and support.
- Leaves of different plants differ in size and shape. The arrangement of veins on the lamina of a leaf is called venation. Two types of venations are seen in plants.
- Flowers are the reproductive organ of a plant.
- The transfer of pollen grains from the anther to the stigma of a flower is called pollination.
- Fertilization takes place soon after pollination; as a result, ovules change into seeds.

Concept Map

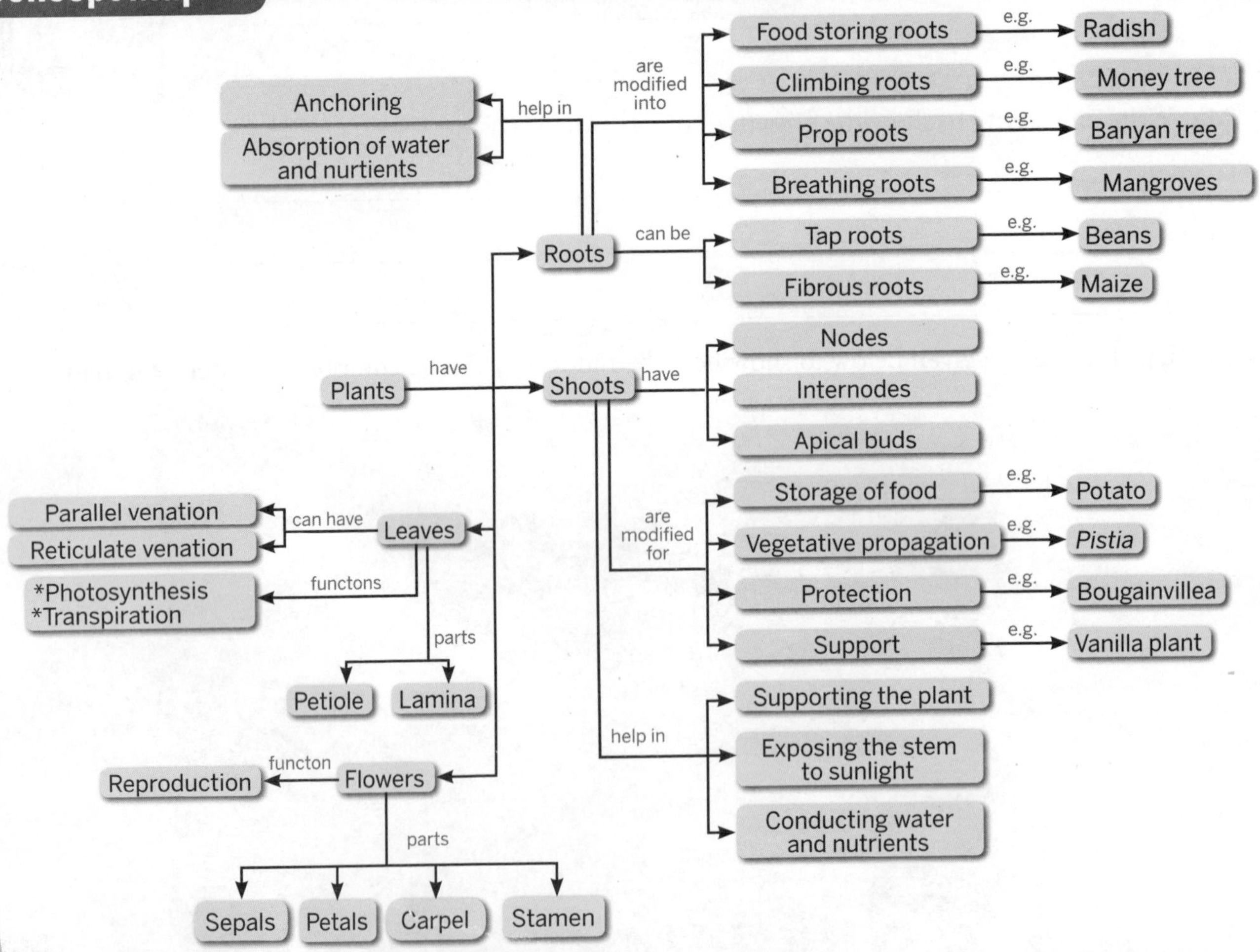

Worksheet 1

Name:

Grade: A B C D

A. Choose the correct option.

1. Roots of which of the following plants grow deep into the soil?
   (a) Wheat (b) Turnip (c) Maize (d) Sugarcane
2. Which of these parts of the plants helps in the absorption of water and nutrients from the soil?
   (a) Flowers (b) Stem (c) Leaves (d) Roots
3. Which of these plants contains prop roots?
   (a) Maize (b) Sweet potato (c) Black pepper (d) Betel
4. Which of the following is the underground modified stem for storing food?
   (a) Creeper (b) Rhizome (c) Climber (d) Tendril
5. Which of the following plants has a weak stem?
   (a) Pumpkin (b) Potato (c) Bougainvillea (d) Rose

B. Match the following.

6. Food-storing roots	(a) Mangroves
7. Prop roots	(b) Pumpkin
8. Creeper	(c) Money plant
9. Climbing roots	(d) Banyan
10. Breathing roots	(e) Radish

C. Give two examples for the following.

11. Climbers ____________________
12. Creepers ____________________
13. Underground stems that store food ____________________
14. Plants with climbing roots ____________________
15. Edible food storing roots ____________________

D. State whether the following statements are true (T) or false (F).

16. Stem is the underground part of a plant.
17. Fibrous roots grow very deep into the soil.
18. The sweet potato plant has fibrous roots.
19. Breathing roots provide mechanical support to the plant.
20. The thorns of bougainvillea plant protect it from predators.

Worksheet 2

Name:

Grade:

A B C D

A. Answer the following questions in a sentence.

1. Name the parts of the plants found below the ground.

2. Name the roots that grow from the main root in the soil.

3. What is the other name for the nodes that have buds from which a new shoot grows?

4. Name the modified structures of bougainvillea that help them from predators.

5. Mention one importance of stem for a plant?

B. Answer the following questions in brief.

6. What constitutes the root system?

7. Write a short note on fibrous roots.

8. What are the functions of the stem?

9. Define vegetative propagation.

10. Distinguish between node and internode.

C. **Answer the following questions in detail.**

11. Distinguish between tap root system and fibrous root system.

____

____

____

12. Label the different parts of a plant given below.

13. Name the category to which the plants shown in the given pictures belong. Write the differences between them.

____

____

____

____

14. Describe the functions of roots.

____

____

____

15. Provide any four examples of the plants that undergo stem modification and mention how these modifications help these plants.

____

____

____

Worksheet 3

Name: ____________

Grade: (A) (B) (C) (D)

A. Choose the correct option.

1. Which of the following leaves has parallel venation?
   (a) Hibiscus (b) Wheat (c) Rose (d) Mango
2. Which of the following parts of the plant manufactures food?
   (a) Leaf (b) Root (c) Flower (d) Fruit
3. The leaves of which of the following plants are modified into leaf spines?
   (a) Glory lily (b) Onion (c) Prickly pear (d) Betel
4. Which of the following carries ovules?
   (a) Ovary (b) Stigma (c) Anther (d) Style
5. ____________ are fine powdery particles that cause fertilization.
   (a) Sepals (b) Petals (c) Pollen grains (d) Stamens

B. Match the following.

6. The brightly coloured part of a flower	(a) Stamen
7. The sticky part that traps pollen grains	(b) Sepal
8. The female reproductive organ of a flower	(c) Stigma
9. The male reproductive organ of a flower	(d) Petal
10. The leaf-like structures that protect the bud	(e) Carpel

C. Give two examples for the following.

11. Plants with parallel venation ____________
12. Seeds that contain only one cotyledon ____________
13. Plants with leaf spines ____________
14. Plants with leaf tendrils ____________
15. Plants with scale leaves ____________

D. Name the following.

16. The stalk of the leaf ____________
17. The outer covering of the seed ____________
18. The swollen base of the pistil that contains the ovules ____________
19. The flat part of the leaf ____________
20. The type of leaf modification seen in the wild pea plant ____________

Worksheet 4

Name:

Grade: (A) (B) (C) (D)

A. Answer the following questions in a sentence.

1. Name different parts of a leaf.

2. How does a plant breathe?

3. How are flowers useful for plants?

4. What is pericarp?

5. How do cotyledons help the developing embryo inside the seed?

B. Answer the following questions in brief.

6. Distinguish between parallel venation and reticulate venation giving examples.

7. Define pollination.

8. Distinguish between pistil and stamen.

9. What are the different functions of the leaves?

10. What are scale leaves? How is it useful to the plants?

C. **Answer the following questions in detail.**

11. Label the different parts of the structure of a flower given below and mention their function.

12. How insects aid in the pollination in plants?

13. Label the parts of a seed given below and write their function.

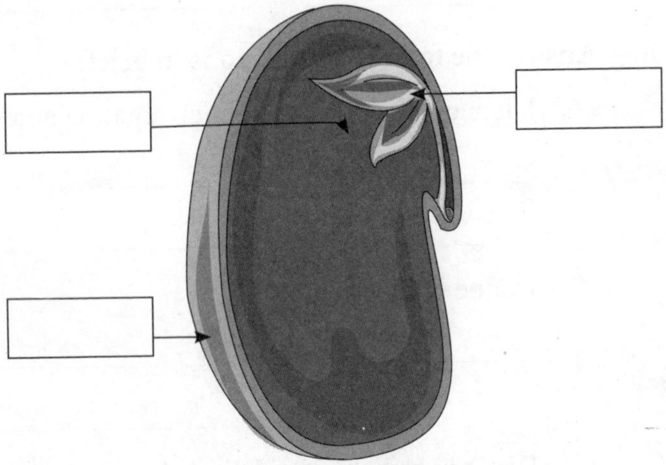

14. Explain any two modifications of leaves.

15. How do you prove that leaves prepare food and store it in the form of starch?

Worksheet 5

Name:

Grade: A B C D

A. Choose the correct option.

1. Which of the following plants contains climbing roots?
   (a) Mangroves (b) Money plant (c) Banyan tree (d) Carrot
2. Which of the following leaves has reticulate venation?
   (a) Wheat (b) Onion (c) Hibiscus (d) Sugarcane
3. Which among the following is not a floral part?
   (a) Filament (b) Petiole (c) Petal (d) Sepal
4. Which of these parts of a flowering plant contains pollen?
   (a) Anther (b) Filament (c) Stigma (d) Style
5. Which of the following is an edible root?
   (a) Potato (b) Radish (c) Mushroom (d) Onion

B. State whether the following statements are true (T) or false (F).

6. The number of sepals and petals are always the same in a flower.
7. All plants contain fibrous roots.
8. The seeds of all the plants contain two cotyledons.
9. Plants with breathing roots grow in marshy regions or salt lakes.
10. Leaves of different plants have same size and shape.

C. Give reasons for the following.

11. A potato left for a long period in the kitchen develops small branches from the small depressions on its surface.

    ____

12. Plants like watermelon grow horizontally on the ground.

    ____

13. White carnation flower becomes red when placed in red coloured water.

    ____

14. A piece of ginger planted in the soil can give rise to a new plant.

    ____

15. Leaves of a glory lily plant are modified into tendrils.

    ____

D. Give two examples for the following.

16. Plants with tap root system ________________
17. Plants with reticulate venation ________________
18. Seeds with two cotyledons ________________
19. Stem tubers ________________
20. Plants with stems or branches modified as thorns ________________

E. Answer the following questions in detail.

21. Name the category of roots to which the plants shown below belong. Write the differences between these two categories of roots.

(a)

(b)

22. Write a note on vegetative propagation.

23. Explain the role of a stem in storage of food, protection and support to the plants.

24. Label the different parts of a leaf and mention their function.

25. How do you prove that roots absorb water and nutrients from the soil?

9 Movement in Animals

Recap of the Chapter

- Animals without a backbone are called invertebrates and animals with a backbone are called vertebrates.
- Birds show two types of flight—flapping and gliding movements.
- Movement in human beings takes place with the help of the skeletal system and the muscular system.
- Bones are the hard parts of the skeleton and cartilage is the soft part of the skeleton.
- One bone is connected to the other with the help of joints for the free movement of the body.
- The joints in the human body are broadly classified on the basis of their movement as immovable joints, slightly movable joints and freely movable joints (ball and socket joint, pivot joint, hinge joint and gliding joint).
- Muscles are of two types—voluntary muscles and involuntary muscles.
- Muscles are connected to bones with the help of band-like tissues called tendons.
- The ends of bones are connected with the help of band-like tissues called ligaments.
- Cartilage protects the bone and also helps it to move.
- X-rays are the rays that can pass through muscles and tissues but not through bones.

Concept Map

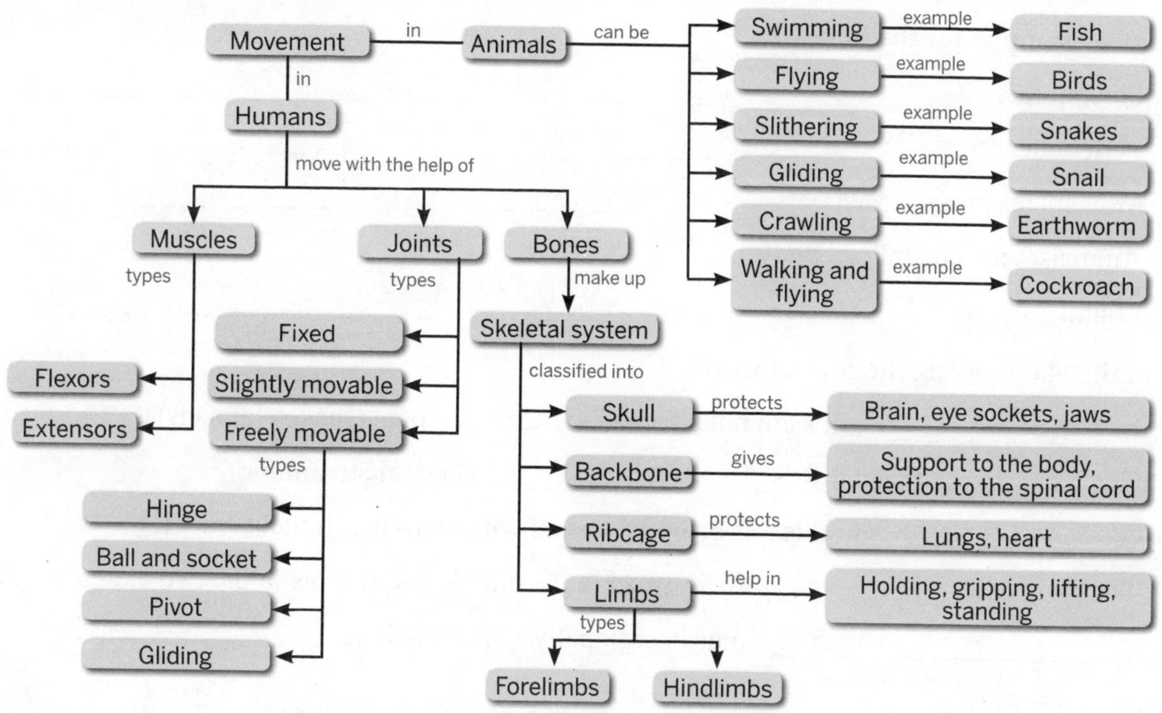

Worksheet 1

Name: ______

Grade: Ⓐ Ⓑ Ⓒ Ⓓ

A. Choose the correct option.

1. Which of the following has a thick muscle foot?
   (a) Snake (b) Earthworm (c) Snail (d) Cockroach
2. Identify the animal that has scales, backbone and ribs.
   (a) Human being (b) Snake (c) Fish (d) Snail
3. What is the shape of the body of a fish?
   (a) Conical (b) Circular (c) Rectangular (d) Streamlined
4. Which of these body parts is used by an earthworm for its movement?
   (a) Tail (b) Wings (c) Fins (d) Setae
5. Which of the following animals does not have bones?
   (a) Snake (b) Bird (c) Earthworm (d) Fish

B. State whether the following statements are true (T) or false (F).

6. All animals do not have bones to support their body movements.
7. Snakes move quickly on rough surfaces.
8. Gliding and soaring are seen in pigeons and parrots.
9. Cockroach has a ribcage to protect its internal organs.
10. Sparrow shows a flapping type of flight.

C. Give two examples for the following.

11. Invertebrates ______________________
12. Vertebrates ______________________
13. Movements in cockroach ______________________
14. Animals with streamlined body ______________________
15. Flightless birds ______________________

D. Fill in the blanks using the correct words.

16. In ______________, ribs are not attached to the chest bone. (snakes/humans)
17. The body shape of the birds is ______________. (spherical/streamline)
18. A ______________ trail is left behind the snail while moving. (mucus/bubble)
19. The body of a ______________ is broad in the middle and narrow at the two ends. (fish/snake)
20. ______________ flap their wings occasionally. (Sparrow/Kites)

Worksheet 2

Name: ____________

Grade: A B C D

A. Answer the following questions in a sentence.

1. What are the two categories of animals based on the presence or absence of a backbone?

____

2. What is a gait?

____

3. Define a swim bladder.

____

4. What are the two types of flights shown by birds?

____

5. What is the role of shell in a snail?

____

B. Answer the following questions in brief.

6. Explain the body structure of a fish.

____

____

7. What are the different body parts used by the cockroach for its movement?

____

____

8. Explain the movement in snails.

____

____

9. What are the special features that help the birds to fly?

____

____

10. What are the uses of fins in fish?

____

____

C. **Answer the following questions in detail.**

11. Differentiate between locomotion and movement.

12. Differentiate between vertebrates and invertebrates with an example.

13. Explain the type of movement in earthworms.

14. Write a note on movement in snake.

15. Label the locomotory organs of a cockroach and write a note on it.

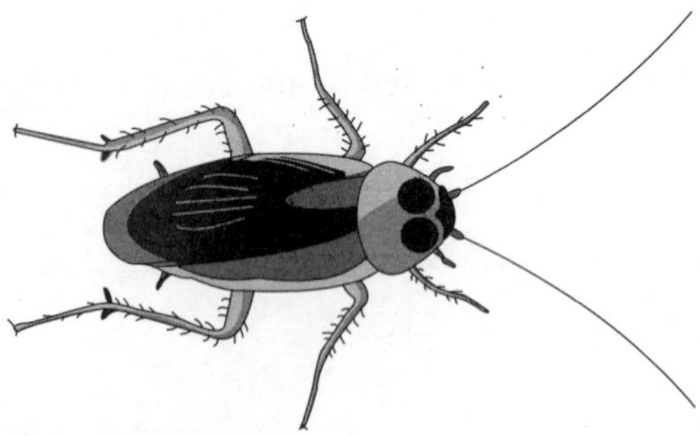

Worksheet 3

Name: ____________

Grade: (A) (B) (C) (D)

A. Choose the correct option.

1. Which of the following connect the ends of the bones together?
   (a) Muscles (b) Ligaments (c) Tendons (d) Cartilage
2. The human spine consists of how many bones?
   (a) 14 (b) 8 (c) 33 (d) 12
3. In which region of the body is ribcage located?
   (a) Abdomen (b) Chest (c) Neck (d) Head
4. How many pairs of ribcage form the floating ribs?
   (a) 8 (b) 6 (c) 4 (d) 2
5. Which of these is connected by tendons at a joint?
   (a) Bone to bone (b) Muscle to muscle (c) Muscle to bone (d) Bone to blood vessel

B. State whether the following statements are true (T) or false (F).

6. The upper part of the skull is made of 14 bones.
7. The skull encloses the brain and protects it from injury.
8. The forelimbs help to bear the weight of the body.
9. The chest bone is also called as sternum.
10. Cartilage is a hard tissue.

C. Give one example for the following.

11. Immovable joint ____________
12. Gliding joint ____________
13. Ball and socket joint ____________
14. Pivot joint ____________
15. Hinge joint ____________

D. Fill in the blanks using the correct words.

16. An ____________ gives an exact image of the fractured bone. (X-ray/laser)
17. The range of movement of a ____________ is backward and forward. (neck/wrist)
18. ____________ jaw cannot be moved up and down. (Upper/Lower)
19. The longest bone in the body is ____________. (femur/stapes)
20. The upper part of an ear is made of ____________. (bone/cartilage)

Worksheet 4

Name: ____________

Grade: Ⓐ Ⓑ Ⓒ Ⓓ

A. Answer the following questions in a sentence.

1. What are joints?

   ____

2. Name the different types of joints in a human body.

   ____

3. Name the two types of muscles.

   ____

4. What are immovable joints and where are they found?

   ____

5. What is a cartilage?

   ____

B. Answer the following questions in brief.

6. Differentiate between skeletal system and skeleton.

   ____

   ____

7. List the main bones in the human skeletal system.

   ____

   ____

8. List the functions of ribcage.

   ____

   ____

9. Write a note on ball and socket joint.

   ____

   ____

10. Give an example of flexors and extensors.

    ____

    ____

C. **Answer the following questions in detail.**

11. List the functions of the backbone.

12. Write a note on the bones of the forelimbs.

13. Differentiate between immovable and slightly movable joints.

14. Explain the working of voluntary muscles.

15. Identify the diagram given below, label it and write a note on the same.

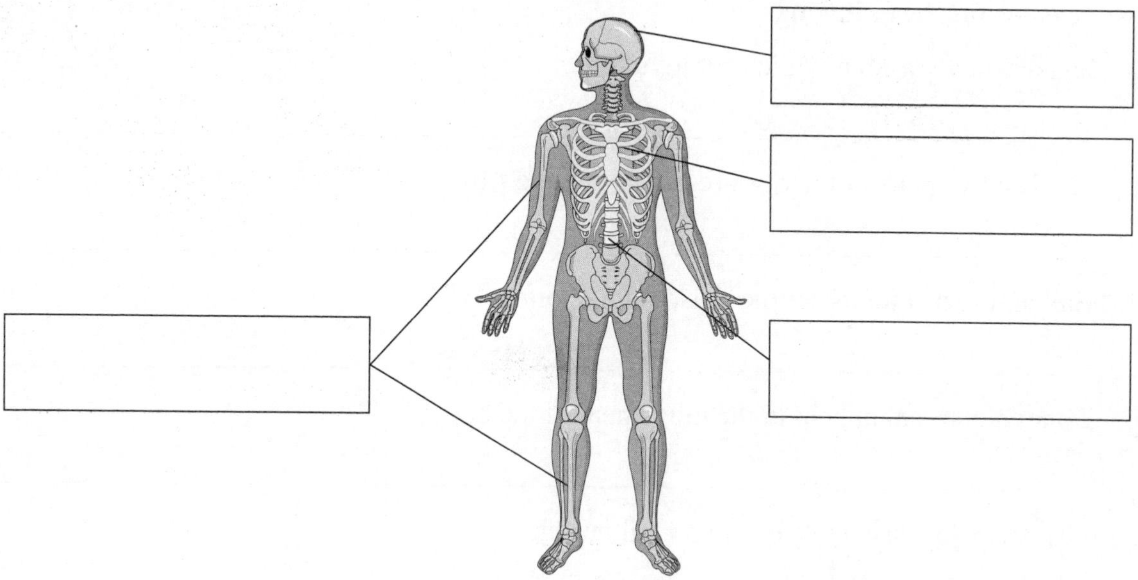

Worksheet 5

Name: ______

Grade: A B C D

A. Choose the correct option.

1. Which of the following protects the heart?
   (a) Skull (b) Ribcage (c) Limbs (d) Backbone
2. What is the rounded structure that a snail carries on its back called?
   (a) Fins (b) Scales (c) Shell (d) Segments
3. Which of these has a long backbone?
   (a) Snake (b) Cockroach (c) Snail (d) Earthworm
4. Which of the following joints can be slightly movable?
   (a) Skull (b) Backbone (c) Hip (d) Ankle
5. Which of the following is an involuntary muscle?
   (a) Biceps (b) Myocardium (c) Triceps (d) Quadriceps

B. Match the following.

6. Movement in earthworm	(a) Flight
7. Movement in snail	(b) Crawling
8. Movement in fish	(c) Slithering
9. Movement in birds	(d) Swimming
10. Movement in snake	(e) Gliding

C. Give reasons for the following.

11. Snakes move fast but not in a straight line.

____

12. The last two pairs of the ribs are called floating ribs.

____

13. Ball and socket joints are the most mobile joints.

____

14. Bones alone cannot help in the movement of the body.

____

15. Fish swim by curving its body and tail muscle.

____

D. Name the following.

16. A joint that allows a movement of 360° in all directions ____________

17. A joint that allows circular, up and down movements ____________

18. A joint that allows movement in only one plane ____________

19. A joint that allows backward and forward movement ____________

20. A joint that does not allow any movement ____________

E. Answer the following questions in detail.

21. Differentiate between hinge joint and gliding joint.

____

____

____

22. What are the modifications seen in birds that help them fly?

____

____

____

23. Write a note on the bones of the hindlimb.

____

____

____

24. Explain the movement of fish in water.

____

____

____

25. Observe the images given below and explain the mechanism.

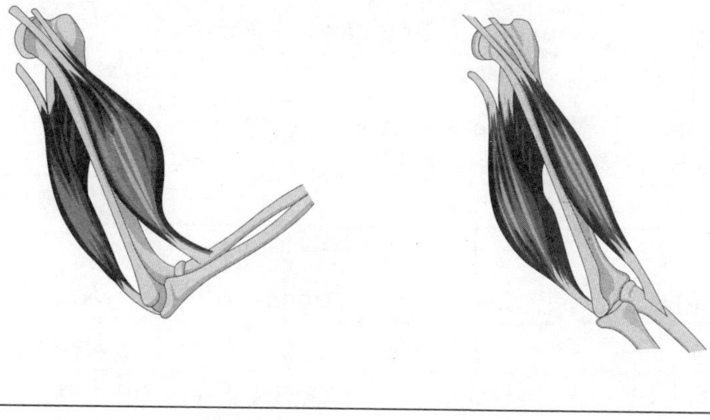

____

____

____

10 Habitat and Adaptation

Recap of the Chapter

- The natural place where an animal lives is called its habitat.
- A habitat consists of living or biotic part and non-living or abiotic part.
- Habitats are broadly classified into land or terrestrial and water or aquatic habitats.
- Terrestrial habitats include grasslands, mountains, different types of forests, deserts and polar regions.
- Aquatic habitats include freshwater, marine, estuaries and coral reefs.
- The features and habits of living organisms that help them to survive in their surroundings are called adaptations.
- Adjustments made by human beings to overcome chronic environmental changes in a short time are called acclimatization.

Concept Map

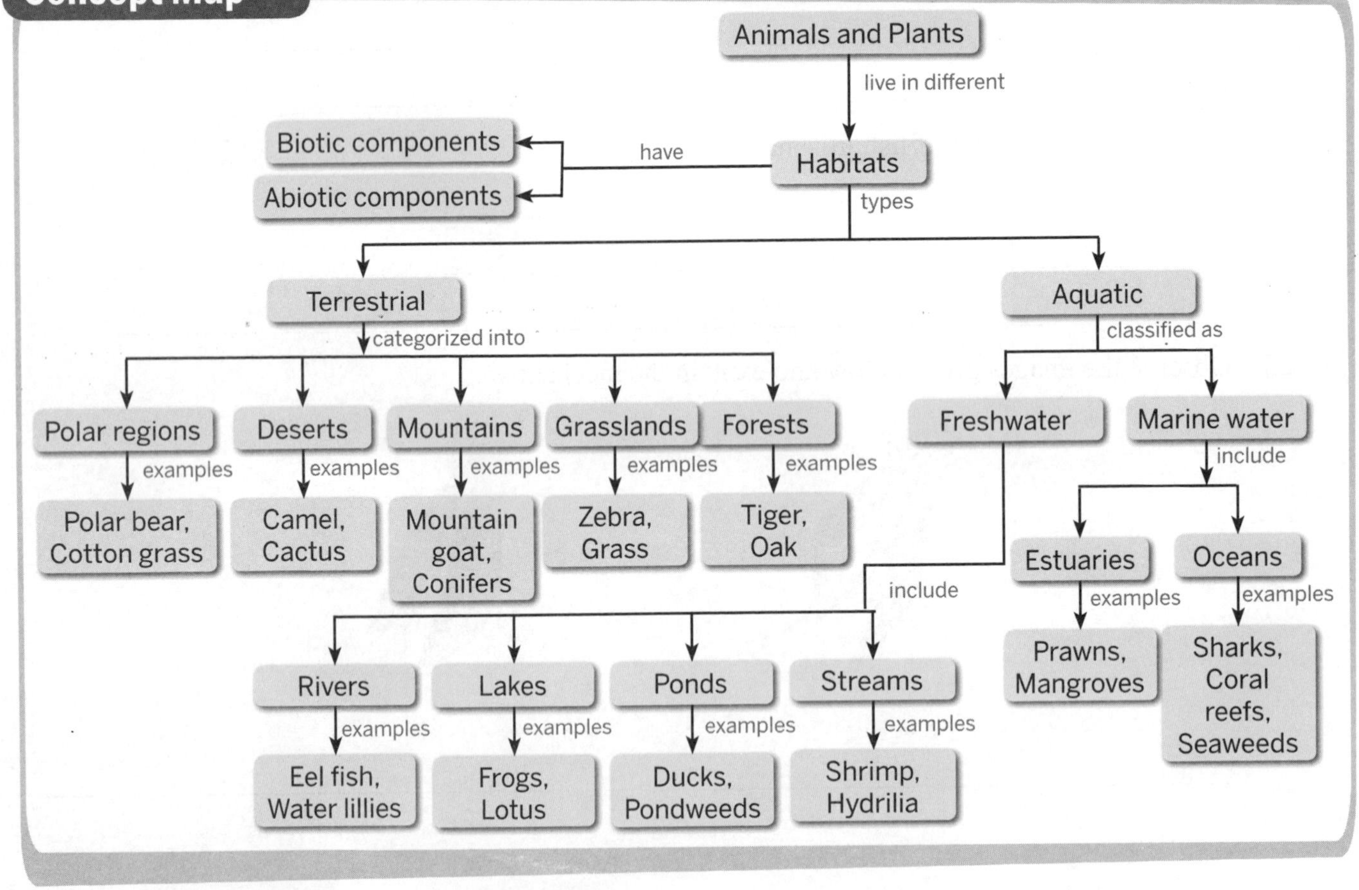

Worksheet 1

Name: ______

Grade: A B C D

A. Choose the correct option.

1. Which of the following is an aquatic habitat?
   (a) Boreal forest (b) Marine (c) Grassland (d) Polar region
2. Which of the following habitats is made up of saltwater?
   (a) Ocean (b) Estuaries (c) River (d) Streams
3. Which of the following is the habitat of a camel?
   (a) Polar region (b) Desert (c) Mountain (d) Forest
4. Which of the following is the habitat for water lily and lotus?
   (a) Freshwater (b) Marine water (c) Coral reef (d) Estuaries
5. Pine trees are found in
   (a) Boreal forest (b) Tropical rainforest (c) Grassland (d) Cold desert

B. State whether the following statements are true (T) or false (F).

6. The term biotic means non-living part of the environment.
7. Terrestrial habitats support life on land.
8. Same kind of plants and animals are found in different habitats.
9. All terrestrial animals can partly live on land and in water.
10. All aquatic plants can float on the surface of water.

C. Name the following.

11. Habitat of giraffe ______
12. Habitat of yak ______
13. Habitat of hornbill ______
14. Habitat of kangaroo rat ______
15. Habitat of penguin ______

D. Fill in the blanks using the correct words.

16. Desert habitats are ______. (cold and wet/hot and dry)
17. Plants and animals of a same region share their ______. (habitat/food)
18. Desert regions do not experience ______. (snowfall/rainfall)
19. Many species of plants and animals are found in ______ region. (equatorial/tropical)
20. The habitat of a duck is ______. (ocean/freshwater)

Worksheet 2

Name:

Grade: A B C D

A. Answer the following questions in a sentence.

1. What is meant by terrestrial habitat?

2. What components constitute a habitat?

3. Name the two broad classifications of habitats?

4. Which habitat is found near the equatorial region?

5. Seals and walruses belong to which habitat?

B. Answer the following questions in brief.

6. Describe mountain habitat briefly.

7. Give one difference between terrestrial and aquatic animals.

8. Name the animals that live in estuaries.

9. Which habitat do dolphins and sharks belong to?

10. Where do turtles and starfish live?

C. **Answer the following questions in detail.**

11. Identify and describe the habitat in the given picture.

12. Describe coral animals with examples.

13. Describe briefly the conditions of tropical, temperate and boreal forest habitats.

14. Give the similarities between the habitat of pine trees and cactus?

15. Identify and describe the given habitat.

Worksheet 3

Name:

Grade:

A B C D

A. Choose the correct option.

1. Which of the following is adapted to live in boreal and temperate forest?
   (a) Black spruce (b) Mosses (c) Lichens (d) All the above
2. Where do grizzly bears live?
   (a) Polar region (b) Boreal forest (c) Desert (d) Mountain
3. Common feature of a cheetah and tiger is
   (a) Camouflaging skin (b) Long tail
   (c) Eating habit (d) Size of head
4. Which of the following is a common practice seen in the animals in polar regions to survive extreme cold weather in winter?
   (a) Aestivate (b) Hibernate (c) Be more active (d) Eat more food
5. Giraffe is adapted to live in which of the following habitats?
   (a) Rainforest (b) Desert (c) Mountain (d) Polar region

B. State whether the following statements are true (T) or false (F).

6. Cheetahs have sharp retractable claws.
7. Grasses have long and strong roots.
8. Fox is an example of a predator.
9. Monkeys do not help in regeneration of forest.
10. Camouflaging does not help animals to escape from predators.

C. Complete the pattern.

11	Marine	:	Saltwater	::	River	:	________
12.	Cacti	:	Xerophyte	::	Hydrilla	:	________
13.	Mosses	:	Boreal forest	::	Tall grasses	:	________
14.	Porcupine	:	________	::	Kangaroo rat	:	Desert
15.	Polar bear	:	Hibernate	::	Wood frog	:	________

D. Name the following.

16. An animal that hibernates ________
17. An animal which uses its tail as fifth limb ________
18. A bird that has strong beak ________
19. An animal which mostly feeds on succulent leaves ________
20. A tree that grows in boreal forest ________

Worksheet 4

Name: ____________

Grade: A B C D

A. Answer the following questions in a sentence.

1. What is the importance of the shape of coniferous trees?

____

2. What type of feet do mountain animals have?

____

3. What is the other name for freshwater plants? Give two examples of freshwater plants.

____

4. What is acclimatization?

____

5. Write the main feature of plants found in deserts.

____

B. Answer the following questions in brief.

6. Describe two common adaptations of animals in desert conditions.

____

____

7. Elephants are known as well diggers. Justify.

____

____

8. How are plants in the lower levels adapted in the rainforest habitat?

____

____

9. Which features should oak, teakwood and rosewood tree adapt to?

____

____

10. What is meant by adaptation?

____

____

C. Answer the following questions in detail.

11. Explain the adaptations of the animal given below.

12. Write three similarities in the adaptations seen in the animals in the given pictures.

13. What are hydrophytes? Explain their adaptive features.

14. Describe the adaptive features of a tiger that help it to survive in grasslands.

15. List four adaptive features of a Bactrian camel.

Worksheet 5

Name:

Grade: A B C D

A. Choose the correct option.

1. Which of the following habitats is very cold for most part of the year?
   (a) Cold desert (b) Freshwater (c) Polar region (d) Marine
2. Snowshoe hares are found in which of the following habitats?
   (a) Boreal forest (b) Desert (c) Marine (d) Mountain
3. Which of the following is home for rattle snakes?
   (a) Desert (b) Ocean (c) River (d) Rainforest
4. Which of the following animals are found in a mountain habitat?
   (a) Yak (b) Llama (c) Mountain goat (d) All of these
5. Which of the following is home for lobsters?
   (a) Rivers (b) Streams (c) Coral reef (d) Waterfalls

B. Match the following.

6. Plant in desert	(a) Waxy coating
7. Leaf of lotus	(b) Fleshy stem
8. Sea grass	(c) Estuaries
9. Planktons	(d) Fixed root
10. Water lily	(e) Marine habitat

C. Give reasons for the following.

11. Sahara is the hottest desert in the world.

    ____

12. Wood frogs are also called frozen frogs.

    ____

13. Horseshoe crabs are found only in estuaries.

    ____

14. Coniferous trees are abundant in temperate forests.

    ____

15. Gobi desert is also known as cold desert.

    ____

D. **Give one example for the following.**

16. Plants that live in desert ______________________
17. An animal found in tropical rainforest ______________________
18. A bird with strong beak ______________________
19. A tall tree found in a temperate forest ______________________
20. A floating water plant ______________________

E. **Answer the following questions in detail.**

21. Explain acclimatization in human beings.

____

____

____

22. Explain briefly the climatic conditions in three types of forests.

____

____

____

23. Explain the adaptations seen in plants found in boreal forests.

____

____

____

____

24. Describe the predator-prey adaptations seen in animals found in grasslands.

____

____

____

____

25. Mention the adaptations that help snakes and rats to survive in hot deserts?

____

____

____

11 Measurement and Motion

Recap of the Chapter

- Measurement is the process of determining the size, amount, etc. of a physical quantity.
- In ancient times, people used their body parts for measurement.
- SI units are the standard set of units for measurement.
- The SI unit of length is metre.
- Length can be measured using a ruler or a measuring tape.
- A flexible and non-stretchable string can be used to measure the length of a curved line.
- A body is said to be in motion when it changes its position with respect to its surroundings.
- If the position of a body does not change with time with respect to its surroundings, then we say the body is at rest.
- There are different types of motion such as translational motion, rotational motion and periodic motion.

Concept Map

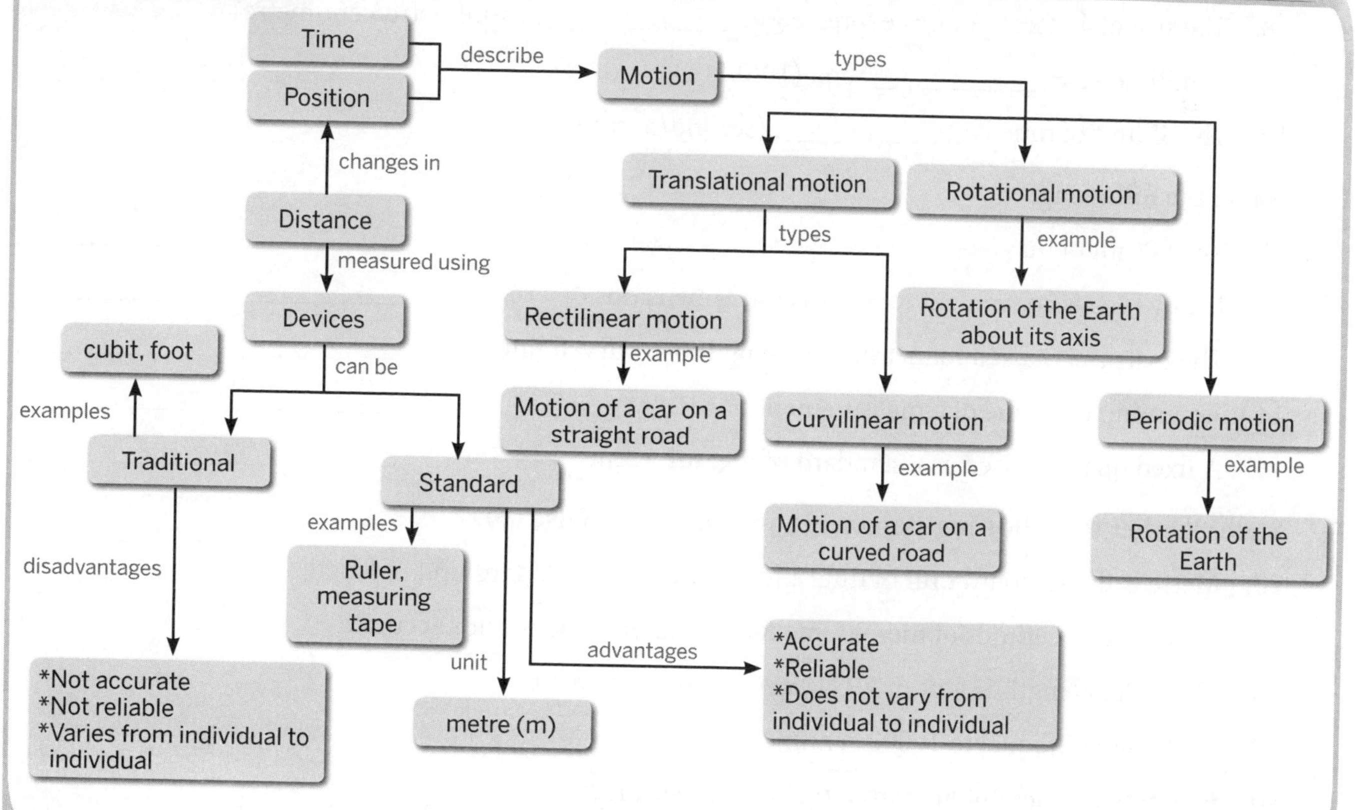

Worksheet 1

Name: ____________

Grade: (A) (B) (C) (D)

A. Choose the correct option.

1. Identify the unit that would be appropriate to measure the length of a plant.
   (a) millimetre (b) centimetre (c) inch (d) metre
2. Ajay's home and school are 5 km apart. Which among the following is the correct representation of this distance in metre?
   (a) 5000 m (b) 50,000 m (c) 5 m (d) 500 m
3. Which of the following is the SI unit of mass?
   (a) Gram (b) Milligram (c) Kilogram (d) Metre
4. How many years make one millennium?
   (a) 10 (b) 100 (c) 50 d) 1000
5. Which among the following is not a unit of length?
   (a) kilometre (b) kilogram (c) metre (d) centimetre

B. Fill in the blanks using the correct words.

6. 1.8 km = ____________ m. (1800/180)
7. The way of assuming the value of a measurement is called ____________. (calculation/estimation)
8. The size of a wheat grain is expressed in ____________. (millimetre/centimetre)
9. 1 millimetre = ____________ cm. (10/0.1)
10. The SI unit of time is ____________. (second/minute)

C. Name the following.

11. The SI unit of mass ____________
12. Unit which is convenient to express the dimensions of a room ____________
13. The instrument used to measure the length of a curved line ____________
14. Assuming the value of a measurement ____________
15. A fixed quantity used as a standard of measurement ____________

D. State whether the following statements are true (T) or false (F).

16. Metre is the standard unit of time and second is the standard unit of length.
17. The ancient methods of measurement are neither standard nor accurate.
18. The unit of length is foot in FPS system of measurement.
19. 1 kilometre is equal to 10000 centimetres.
20. Estimation is helpful in many situations in our life.

Worksheet 2

Name: ______

Grade: A B C D

A. Answer the following questions in a sentence.

1. Meera places one end of a stick at 100 cm and the other end at 80 cm, on a metre scale. What is the length of the stick? Also, write the calculations involved.

2. How many equal divisions of millimetres are there in two centimetres?

3. A cyclist travels a distance of 2500 m in a certain time. What would be the distance covered by the cyclist in kilometres?

4. Arrange the units-metre, centimetre, kilometre and millimetre in ascending order.

5. A reading in a test tube shows 125 cm. What is the reading in millimetres?

B. Answer the following questions in brief.

6. Briefly explain the need of using SI units for measuring the objects.

7. Tabulate the units of length, mass and time in CGS and FPS systems of unit.

8. How can we measure the length of a curved line?

9. What is meant by estimation?

10. Write a short note on measurement.

C. **Answer the following questions in detail.**

11. Explain the correct method of measuring the length of an object using a ruler.

12. How can we measure the length of a line segment using a divider and a scale?

13. What is meant by International System of Units?

14. Write a note on the instruments used to measure the length.

15. Find the length of the wooden blocks that are given below.

(a)

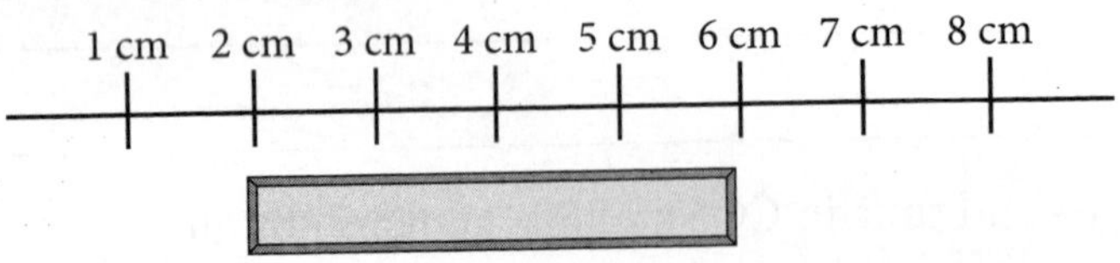

(b)

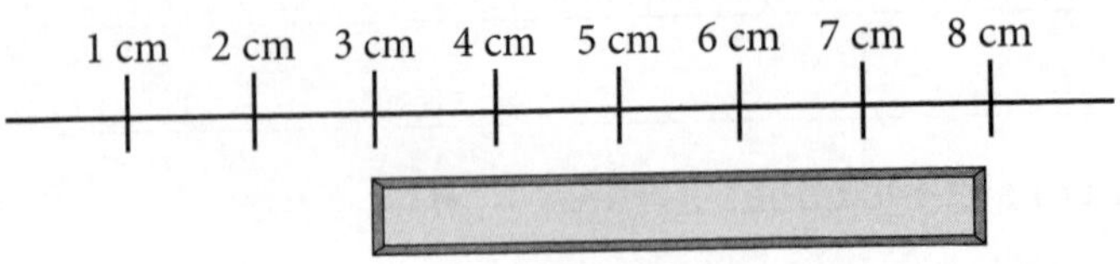

(c)

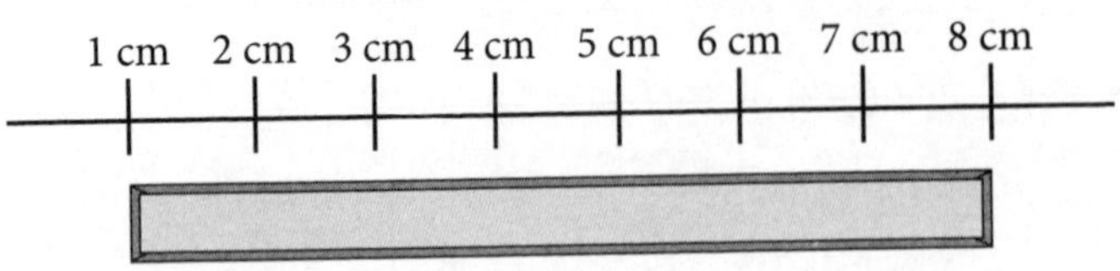

(d)

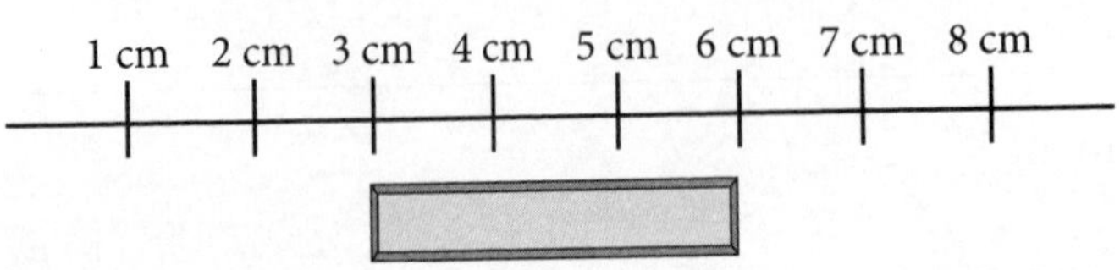

Worksheet 3

Name: ____________

Grade: A B C D

A. Choose the correct option.

1. Which of the following statements is true?
   i. Pendulum of a clock shows periodic motion.
   ii. Pendulum of a clock exhibits oscillatory motion.
   (a) Only i (b) Only ii (c) Both i and ii (d) Neither i nor ii
2. What type of motion is shown by the spinning wheel?
   (a) Rotational (b) Translational (c) Both (a) and (b) (d) Neither (a) nor (b)
3. Which of the following is an example of periodic motion?
   (a) Motion of Earth (b) Motion of cockroach (c) Motion of car (d) None of these
4. Which of these is an example of rectilinear motion?
   (a) Marching of soldiers (b) Rotation of Earth (c) Spinning of wheel (d) Motion of a swing
5. Which of the following undergo rotational and translational motion?
   (a) Wheels of a moving car (b) Motion of pendulum
   (c) A moving cockroach (d) None of these

B. Fill in the blanks using the correct words.

6. A cockroach shows ____________ motion. (random/periodic)
7. A body in/at ____________ changes its position with time. (motion/rest)
8. Wheels of a bicycle show ____________ motion. (rotational/curvilinear)
9. Pendulum of a clock shows ____________ motion. (oscillatory/curvilinear)
10. If the position of the body does not change with respect to surroundings, then we can say that the body is ____________. (at rest/in motion)

C. Name the following.

11. A type of translational motion ____________
12. The type of motion be observed when a car is moving on a straight road ____________
13. The kind of motion exhibited by the spinning top ____________
14. Motion of a truck on a straight road is an example of ____________
15. The type of motion in which an object moves such that the distance from a fixed point remains the same ____________

D. State whether the following statements are true (T) or false (F).

16. The nature of motion of a boy sitting in a moving car is different from the motion of a boy sitting in a roller coaster.
17. When a car is moving, its position does not change with time with respect to its surroundings.
18. The motion of a stone falling down is an example for curvilinear motion.
19. Rotation and revolution of Earth exhibit two different types of motion.
20. A group of police men marching during guard of honour depict periodic motion.

Worksheet 4

Name:

Grade: (A) (B) (C) (D)

A. Answer the following questions in a sentence.

1. When can we say that an object is in motion?

2. Write the condition for an object to be at rest.

3. Which type of motion is responsible for producing sound in stringed musical instruments?

4. Write the relationship between motion and time.

5. Give an example of rotational motion from your daily life.

B. Answer the following questions in brief.

6. Explain curvilinear motion with an example.

7. Earth moving around the sun exhibits both circular and periodic motion. Justify.

8. What is meant by oscillatory motion?

9. Write a short note on rotational motion.

10. How are motion and rest different?

C. Answer the following questions in detail.

11. Explain the two types of translational motion.

____

____

____

12. Write one instance each for the following and justify your answer.
    (a) Rotatory as well as translatory motion (b) Rotational as well as periodic motion

____

____

____

13. Identify the motion from the picture given below and explain it.

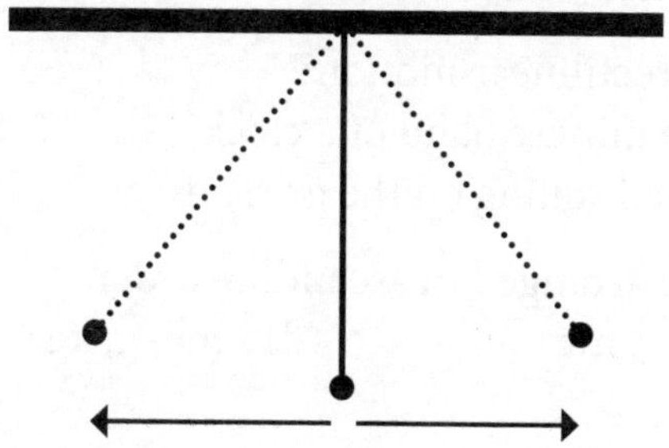

____

____

____

14. Explain different types of motion.

____

____

____

15. How are rotational and translational motions different? Give two examples of each.

____

____

____

Worksheet 5

Name:

Grade: A B C D

A. Choose the correct option.

1. Which of the following is correct with respect to the motion of wheels of a car moving on a straight road and motion of a ceiling fan?
   (a) Both undergo rotatory motion (b) Both undergo rectilinear motion
   (c) Both undergo oscillatory motion (d) Both undergo rotatory motion only

2. Which of the following is the most convenient unit for measuring the thickness of a coin?
   (a) Millimetre (b) Centimetre (c) Metre (d) Kilometre

3. Pick the odd one out.
   (a) Motion of the diaphragm of a drum (b) Motion of a simple pendulum
   (c) Motion of strings of musical instruments (d) Motion of a football

4. Which among the following examples does not exhibit rectilinear motion?
   (a) The wheels of a moving bicycle (b) The minute hand of a clock
   (c) An athlete running on a straight track (d) A ball rolling on the ground

5. Identify the correct order in which the units of mass are arranged in ascending order.
   (a) g, mg, kg (b) kg, mg, g (c) kg, g, mg (d) mg, g, kg

B. Match the following.

6. Dimensions of a classroom	(a) Millimetre
7. Length of the textbook	(b) Metre
8. Distance between Mumbai and Delhi	(c) Smaller than a millimetre
9. Thickness of a 2-rupee coin	(d) Centimetre
10 Screw gauge	(e) Kilometre

C. Give reasons for the following.

11. Traditional methods of measurement are not reliable compared to standard methods of measurement.

____

12. Can a body have rectilinear motion and rotational motion at the same time?

____

13. Generally, the distance between two cities are not measured in centimetres.

____

14. A flexible and non-stretchable string or thread is used to measure the length of a curved line.

____

15. Every translational motion is not rectilinear motion.

____

D. Name the following.

16. The SI unit of time ______
17. The standard set of units for measurement ______
18. The motion of an object repeated at regular intervals of time ______
19. The SI unit of length ______
20. The unit of length used in Vedic times in India ______

E. Answer the following questions in detail.

21. Observe the following pictures and identify the types of motions involved. Also, write their features.

(a)

(b)

22. Skipping is a favourite exercise and pastime for many. What type of motion does it represent? Explain its features.

23. Motion of the hands of a clock and moving pendulum of a clock exhibit different kinds of motion. Comment.

24. Explain the combination of motions that we can observe in the following cases.
(a) The wheels of a moving car (b) Movement of a ball delivered by a bowler

25. Explain the following motions and give one example for each.
(a) Oscillatory motion (b) Periodic motion (c) Rotational motion

12 Electricity and Circuits

Recap of the Chapter

- Electric cell is a device that converts chemical energy into electrical energy.
- The cells that convert solar energy into electrical energy are called solar cells.
- The path through which electric current flows between the two terminals of an electric cell is called electric circuit.
- An electric circuit can be kept closed or open using a switch.
- Electric bulb is a device that produces light when electricity is passed through it.
- A torch is a simple electric circuit containing cells, a bulb and a switch.
- Materials that allow the passage of electricity through them are called conductors.
- Materials that do not allow the passage of electricity through them are called insulators.

Concept Map

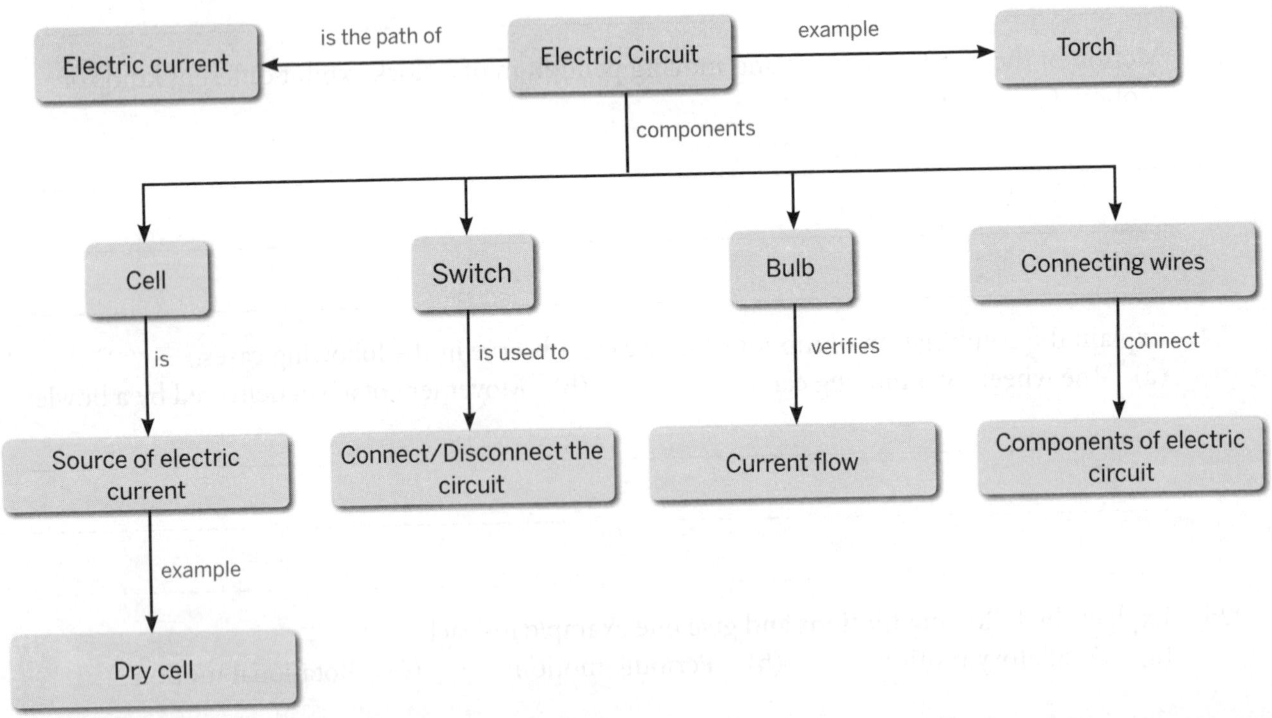

Worksheet 1

Name: ____________

Grade: A B C D

A. Choose the correct option.

1. Identify the device that does not need electricity to work.
   (a) Electric fan (b) Refrigerator (c) Pressure cooker (d) Electric iron
2. Which among the following devices converts chemical energy into electrical energy?
   (a) Electric bulb (b) Switch (c) Electric cell (d) Resistor
3. Which of the following devices helps to keep the circuit closed or open?
   (a) Electric cell (b) Resistor (c) Electric bulb (d) Switch
4. Which is the device that is capable of converting solar energy into electrical energy?
   (a) Solar cell (b) Electric cell (b) Switch (d) Resistor
5. Which among the following devices does not use button cells?
   (a) Wristwatch (b) Digital thermometer (b) Radio (d) LED flashlights

B. Fill in the blanks using the correct words.

6. Electric current flows through a circuit when it is ______________. (open/close)
7. The cells used in a torch are ______________. (dry cells/solar cells)
8. The path through which electric current flows between the two terminals of an electric cell is known as an ______________. (electric cell /electric circuit)
9. When you turn the switch, ______________ the circuit is closed. (on/off)
10. AAA cell is an example of ______________. (button cell/dry cell)

C. Name the following.

11. The part of the torch responsible for providing current ______________
12. The metal used as a container in a dry cell inside which there is a layer of ammonium chloride paste ______________
13. The closed path that allows the flow of electricity ______________
14. The diagram drawn using the symbols of components of the circuit ______________
15. The component of a circuit diagram that can be represented with the symbol ______________

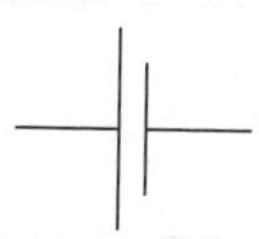

D. State whether the following statements are true (T) or false (F).

16. In torches, the current is provided by electric cells.
17. Electric cell is a device that converts electrical energy into chemical energy.
18. An electric cell has only one terminal.
19. The cells that are capable of converting solar energy into electrical energy are called solar cells.
20. An electric circuit can be kept closed or open using a device called switch.

Worksheet 2

Name:

Grade:

Ⓐ Ⓑ Ⓒ Ⓓ

A. Answer the following questions in a sentence.

1. What is meant by a solar panel?

____

2. What are the main components of a simple electric circuit?

____

3. How many terminals are there in an electric cell? What are they?

____

4. Name three electronic devices that use button cells.

____

5. Write the function of a switch in an electric circuit.

____

B. Answer the following questions in brief.

6. How does a cell produce electric current?

____

____

7. When can we call a circuit as closed circuit?

____

____

8. What is meant by an electric cell? Explain how it produces electricity.

____

____

9. What are the applications of solar cells?

____

____

10. Write a note on the terminal parts of a cell.

____

____

C. Answer the following questions in detail.

11. How do solar panels produce electricity?

12. Write a note on dry cells.

13. Write the applications of the following.
    (a) Dry cells
    (b) Button cells

14. Write the advantages of solar cells over dry cells. Write any one limitation of solar cell.

15. Draw a circuit diagram using the following components. Explain the functions of connecting wires and cell.

(a) 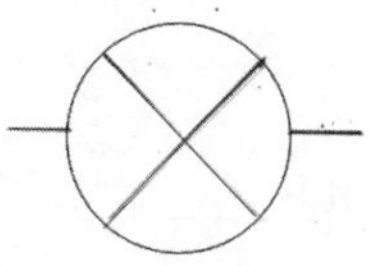

Bulb

(b) 

Switch

(c) 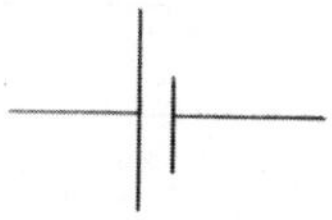

Cell

(d)

Connecting wire

Worksheet 3

Name:

Grade: A B C D

A. Choose the correct option.

1. Which part of the torch produces light when it is switched on?
   (a) Bulb (b) Cell (c) Metal spring (d) Plastic casing
2. Identify the metal that is used to make the filament of a bulb.
   (a) Copper (b) Zinc (c) Tungsten (d) Aluminium
3. Who invented the electric bulb?
   (a) Isaac Newton (b) Thomas Alva Edison
   (c) Archimedes (d) Michael Faraday
4. Which among the following is not a conductor of electricity?
   (a) Rubber (b) Copper (c) Iron (d) Aluminium
5. Materials that do not allow electricity to flow through them are
   (a) conductors (b) semiconductors (c) copper wires (d) insulators

B. State whether the following statements are true (T) or false (F).

6. A fused bulb can produce light, when connected to four cells.
7. Electric current flows through metals.
8. The part of an electric bulb which produces light is called filament.
9. Plastic is a good conductor of electricity.
10. Water is not used for extinguishing fire caused due to electrical short circuits.

C. Name the following.

11. The coil inside the bulb that glows when electricity is passed through it ______
12. Materials that do not allow electric current to flow through them ______
13. The part of a torch that supplies electric current ______
14. The opposition offered by a conductor to the flow of electric current ______
15. The materials that allow the passage of electricity through them ______

D. Fill in the blanks using the correct words.

16. When you turn the switch ______, the circuit is open and no current flows through the circuit. (OFF/ON)
17. A diagram using symbols of the components of a circuit is called a ______. (circuit diagram/electric cell)
18. Filament of a bulb is usually made of a metal called ______. (tungsten/nichrome)
19. The bulb becomes ______ when an electric current passes through it. (hot/cool)
20. A glass rod is a/an ______ of electricity. (conductor/insulator)

Worksheet 4

Name:

Grade:

A B C D

A. Answer the following questions in a sentence.

1. What is an electric bulb?

2. Why is argon used to fill an electric bulb?

3. Why can we not use rubber or plastic as connecting wires in electrical appliances?

4. What will happen if we plug many sockets to a single plug?

5. When does an electric cell stop producing electricity?

B. Answer the following questions in brief.

6. What is meant by fused bulbs?

7. How can we differentiate between insulators and conductors of electricity?

8. Explain why switch is considered as a safety measure.

9. Why should we keep electrical appliances away from water?

10. We should take extreme care while using electricity. Comment.

C. Answer the following questions in detail.

11. Explain the safety precautions to be taken while dealing with electricity.

12. Elaborate the major parts of a torch.

13. Describe an activity which shows that an electric bulb produces heat when electricity is passed through it.

14. Explain how bulb acts as a light-producing component in a torch.

15. Label the parts of the bulb and explain the role of a filament.

Worksheet 5

Name:

Grade: A B C D

A. Choose the correct option.

1. Identify the direction of flow of electric current through an electric circuit.
   (a) From the positive terminal to the negative terminal of the cell
   (b) From the negative terminal to the positive terminal of the cell
   (c) From the positive terminal to the positive terminal of the cell
   (d) From the negative terminal to the negative terminal of the cell
2. Which of the following gases is used to fill inside an electric bulb?
   (a) Helium (b) Neon (c) Oxygen (d) Argon
3. Zinc container, ammonium chloride paste and carbon rod covered with a metal cap—these are the main components in
   (a) button cell (b) electric bulb (c) switch (d) dry cell
4. Which among the following is not a conductor of electricity?
   (a) Steel spoon (b) Human body (c) Water (d) Plastic bag
5. Which among the following converts energy from the sun into electrical energy?
   (a) Dry cell (b) Solar cell (c) Switch (d) Button cell

B. State whether the following statements are true (T) or false (F).

6. When chemicals in an electric cell are used up, the electric cell stops producing electricity.
7. When we connect more number of cells, a fused bulb can produce light.
8. Human body is not a conductor of electricity.
9. Metals are good conductors of electricity.
10. The part of an electric bulb which produces light is known as filament.

C. Give reasons for the following.

11. Electric current can only flow through a closed circuit.

    ____

12. Tungsten is preferred to be used as a filament in bulbs.

    ____

13. Always wear rubber sandals while handling electrical appliances.

    ____

14. Make sure that all the cords of electrical appliances are unplugged when not in use.

    ____

15. Human body is a fairly good conductor of electricity.

    ____

D. Name the following.

16. The broken circuit that does not let current to pass through it ____________
17. The device that converts chemical energy into electrical energy ____________
18. Cell that converts solar energy into electrical energy ____________
19. Electric bulb with a broken filament ____________
20. The device used to open or close an electric circuit ____________

E. Answer the following questions in detail.

21. Explain the major components of an electric circuit.

____

____

____

22. Differentiate between electric and solar cells.

____

____

____

23. Explain the working of a torch.

____

____

____

24. Describe an activity to show the conductivity of liquids.

____

____

____

25. Label the internal parts of a torch given below. Write the function of dry cell in a torch.

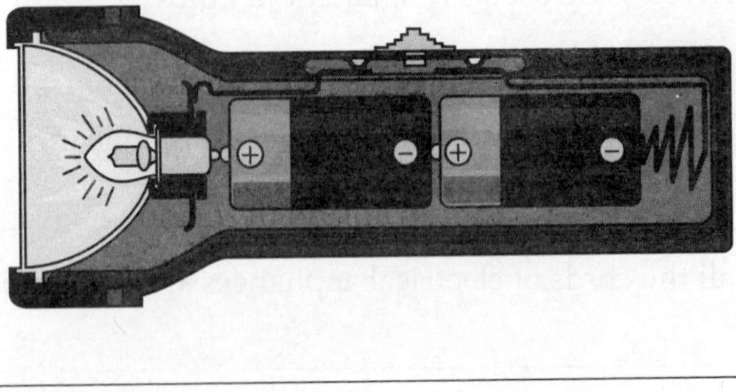

____

____

____

13 Magnets

Recap of the Chapter

- A magnet obtained naturally from the Earth is called a natural magnet.
- An artificial magnet is made of magnetic materials using artificial methods.
- Materials that are attracted towards a magnet are called magnetic materials.
- Materials that are not attracted towards a magnet are called non-magnetic materials.
- The ends of a magnet where the magnetic properties are stronger are called the poles of the magnet.
- The end of a magnet pointing towards North, when suspended freely is called the North Pole of the magnet.
- The end of a magnet pointing towards South, when suspended freely is called the South Pole of the magnet.
- Unlike poles of magnets always attract each other.
- Like poles of magnets always repel each other.
- Magnetic poles can never be isolated.
- A compass is a device used to find directions.
- The process of making a magnet from a magnetic material is called magnetization.
- Magnets have to be preserved to prevent the loss of magnetic properties.
- The process by which a magnet loses its magnetic properties is called demagnetization.

Concept Map

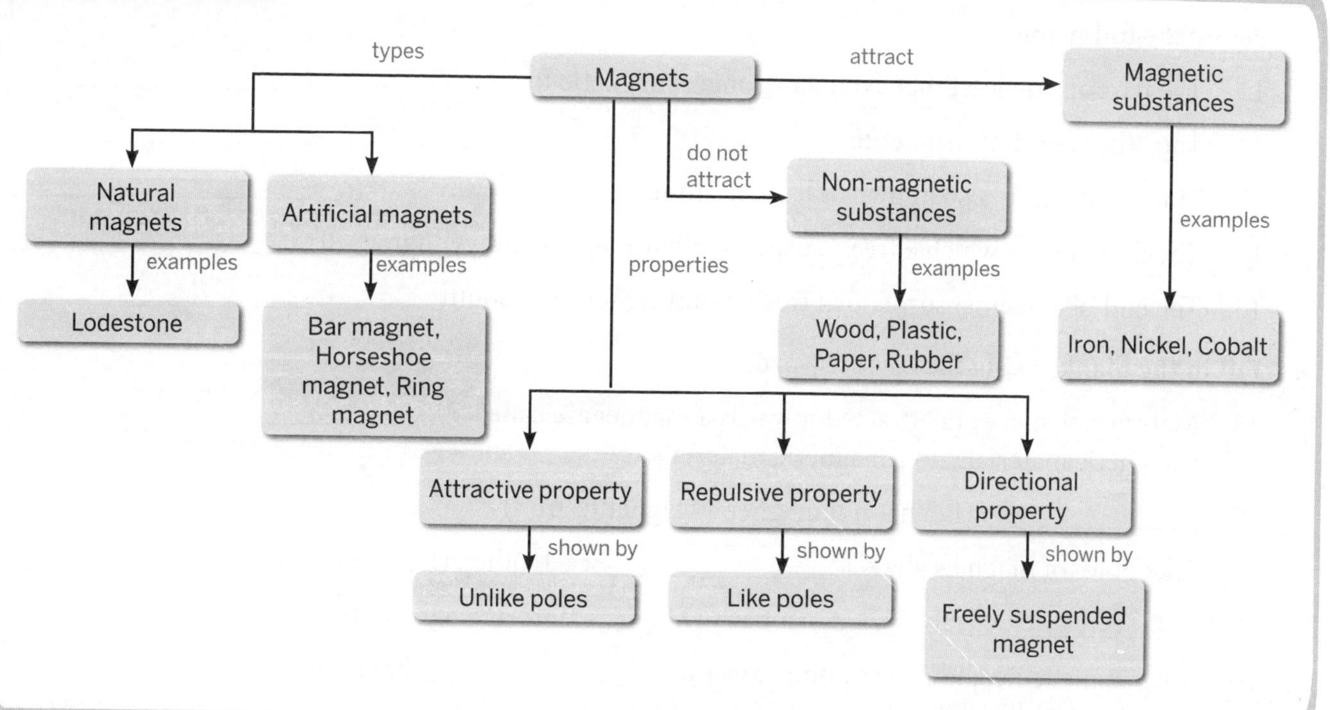

Worksheet 1

Name: ____________

Grade: A B C D

A. Choose the correct option.

1. Which of the following is a magnetic material?
(a) Silicon (b) Iron (c) Carbon (d) Zinc
2. Wood, plastic and rubber are examples of
(a) magnetic materials (b) non-magnetic materials
(c) metals (d) conductors
3. Which of the following devices does not use a magnet?
(a) Loudspeaker (b) Electric cranes (c) Radio (d) Mercury lamp
4. How many poles does a magnet have?
(a) 2 (b) 1 (c) 4 (d) 8
5. Magnets are stored in
(a) wooden boxes (b) magnetic keepers (c) plastic boxes (d) None of these

B. State whether the following statements are true (T) or false (F).

6. Bar magnet is an example of natural magnet.
7. Lodestone is a natural magnet.
8. Cobalt and Nickel are examples of magnetic materials.
9. When we suspend a magnet freely, the end pointing towards North is called South Pole.
10. Unlike poles of magnets always attract each other.

C. Name the following.

11. The part of a magnet that exhibits stronger magnetic force ____________
12. The other name of magnetite ____________
13. This substance is also known as leading stone ____________
14. The direction in which a freely suspended bar magnet always align itself ____________
15. The end of a magnet suspended freely pointing towards South ____________

D. Fill in the blanks using the correct words.

16. Materials that are not attracted towards a magnet are called ____________. (magnetic materials/non-magnetic materials)
17. ____________ is a natural magnet. (Magnetite/Iron)
18. Like poles of magnets always ____________ each other. (repel/attract)
19. ____________ always exist in pairs.(Magnetic forces/Magnetic poles)
20. The magnetic needle always points towards ____________ direction. (North-South/East-West)

Worksheet 2

Name: ____________

Grade: (A) (B) (C) (D)

A. Answer the following questions in a sentence.

1. Which English alphabets are used to represent North and South poles of a magnet?

____

2. What is meant by magnetic poles?

____

3. What are natural magnets?

____

4. What do you mean by repulsive property of magnets?

____

5. What are magnetic materials?

____

B. Answer the following questions in brief.

6. How can we identify the North pole of a bar magnet?

____

____

7. What is a magnetite?

____

____

8. Write an activity to show the repulsive property of magnets.

____

____

9. What do you mean by directional property of magnets?

____

____

10. Write a note on artificial magnets.

____

____

C. Answer the following questions in detail.

11. Explain an activity to find out the poles of a magnet.

12. Explain the fundamental laws of magnetism.

13. Write a note on the discovery of magnets.

14. Explain an activity to demonstrate that the property of attraction is more at the two ends of a magnet.

15. Look at the pictures given below and list the objects that get attracted to a magnet. Write a note on magnetic and non-magnetic materials.

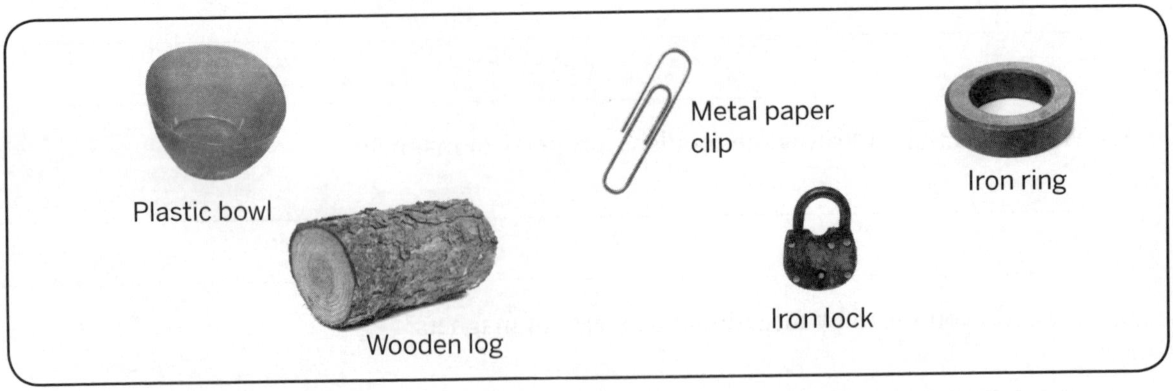

Worksheet 3

Name: ____________

Grade: Ⓐ Ⓑ Ⓒ Ⓓ

A. Choose the correct option.

1. In which direction does the magnetic needle in a compass always point?
   (a) North–South (b) South–North (c) North (d) South
2. Which of the following is a method of making artificial magnets?
   (a) Single-touch (b) Demagnetization (c) Single keeper (d) Double keeper
3. Which of the following magnets is used in an electric bell?
   (a) Bar magnet (b) Cylindrical magnet (c) Horseshoe magnet (d) Electromagnet
4. Which of these will get attracted towards a magnet?
   (a) Iron filings (b) Another magnet (c) Both (a) and (b) (d) Neither (a) nor (b)
5. Which among the following methods is used to make a strong magnet?
   (a) Single-touch method (b) Double-touch method
   (c) Stroke method (d) Electrical method

B. State whether the following statements are true (T) or false (F).

6. It is possible to make a magnet with only one pole.
7. A bar magnet cannot be used to separate a mixture of iron filings and sand.
8. A magnetic compass is a device used to find directions.
9. Earth behaves as a huge magnet whose magnetic North pole is at the geographic South pole.
10. All conductors of electricity exhibit magnetic properties.

C. Name the following.

11. The principle on which a magnetic compass works ____________
12. The device used to find the direction ____________
13. A magnet obtained naturally ____________
14. The device where strong magnets are used ____________
15. A natural stone that has magnetic properties ____________

D. Fill in the blanks using the correct words.

16. The method of magnetizing a material by rubbing with a magnet is called ____________ method. (single-touch/double-touch)
17. Earth acts as a magnet whose magnetic north pole is at the geographic ____________ pole. (North/South)
18. Pieces of iron kept at the ends of a magnet to preserve the magnetism of the magnets are called ____________. (magnetic poles/magnetic keepers)
19. When the south pole of a bar magnet is placed near the north pole of another magnet, they ____________ each other. (attract/repel)
20. The process by which a magnet loses its magnetic properties is called ____________. (demagnetization/magnetization)

Worksheet 4

Name:

Grade:

A B C D

A. Answer the following questions in a sentence.

1. How are bar magnets stored?

____

2. Where are the poles of the Earth's magnet located geographically?

____

3. Name two different stroke methods that are used to make magnets from magnetic materials.

____

4. What is the nature of the magnets made by stroke methods?

____

5. How can a horseshoe magnet be preserved?

____

B. Answer the following questions in brief.

6. What is meant by demagnetization?

____

____

7. Briefly explain the electrical method to make strong magnets.

____

____

8. How a magnet loses its magnetism?

____

____

9. What do you mean by magnetization?

____

____

10. Explain the touch methods of making magnets using a diagram.

____

____

C. **Answer the following questions in detail.**

11. What is Oersted experiment?

12. Write a note on magnetic compass.

13. Earth shows magnetic properties. Do you agree or disagree? Justify your answer.

14. Write an activity for making a magnet using the single-touch method.

15. Mark the polarities of the magnetised iron bar given below. Explain the other touch method used to make a magnet and draw its diagram.

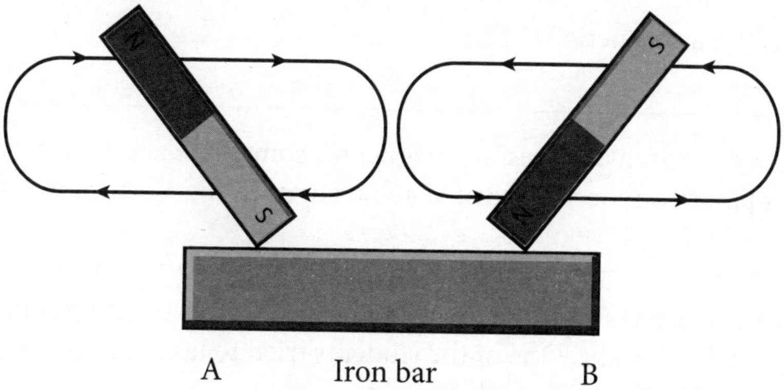

Name:

Grade: A B C D

Worksheet 5

A. Choose the correct option.

1. Which among the following is a magnetic material?
   (a) Wood (b) Plastic (c) Cobalt (d) Rubber
2. A magnet obtained naturally from the Earth is called a
   (a) natural magnet (b) artificial magnet (c) magnetization (d) demagnetization
3. The process of making a magnet from a magnetic material is called
   (a) natural magnet (b) artificial magnet (c) magnetization (d) demagnetization
4. Identify the natural magnet from the following.
   (a) Bar magnet (b) Magnetite (c) Ring magnet (d) Disc magnet
5. The process by which a magnet loses its magnetic properties is called
   (a) natural magnet (b) artificial magnet (c) magnetization (d) demagnetization

B. Match the following.

6. Iron	(a) Natural magnet
7. Aluminium	(b) Magnetic material
8. Lodestone	(c) The ends of a magnet
9. Magnetic poles	(d) Used to find direction
10. Magnetic compass	(e) Non-magnetic material

C. Give reasons for the following.

11. Magnets should be preserved with magnetic keepers.

___

12. A razor blade is brought in contact with a horseshoe magnet for some time and then taken away from the magnet. When it is placed on the surface of water, it aligns in the North-South direction.

___

13. Some students were fixing different pictures on their school bulletin board by using stainless steel pins. Most of the pins fell down by mistake. One of the students tried to take the pins using a magnet. The student could not collect the pins with magnets.

___

14. Neena saw a bar magnet that is not labelled as North and South poles on it. But, she easily found the poles of the magnet.

___

15. It is advised to keep magnets away from wristwatches, computers and mobiles.

___

D. Give one example for the following.

16. Natural magnet
17. Magnetic material
18. Demagnetization
19. Non-magnetic material
20. Artificial magnet

E. Answer the following questions in detail.

21. Explain the uses of magnets.

22. Write an activity to show the attractive and repulsive property of magnets.

23. How can we make a magnet using the single-touch method?

24. Magnetic poles can never be isolated. Why?

25. What are attractive and repulsive properties of a magnet? In which of the following cases do the magnets exhibit attractive and repulsive properties?

Case 1

Case 2

Case 3

Case 4

14 Light, Shadows and Reflection

Recap of the Chapter

- Objects that give out light are called sources of light.
- The objects that emit their own light are called luminous objects.
- The objects that do not emit their own light are called non-luminous objects.
- Light always travels in a straight line. This property of light is called rectilinear propagation.
- When we consider a single and a very narrow path of light, it is usually called as a light ray.
- A light beam consists of a number of light rays which can be parallel to each other.
- The materials that allow light to pass through them completely are called transparent materials.
- The materials that allow light to pass through them partially are called translucent materials.
- The materials that do not allow light to pass through them are called opaque materials.
- The dark area formed when the light is blocked by any opaque object kept in its path is called a shadow.
- The phenomenon by which the light rays falling on a surface bounce back completely or partially is called reflection of light.
- A pinhole camera is a simple camera that consists of a lightproof box with a small hole on one side and a screen on the opposite side.

Concept Map

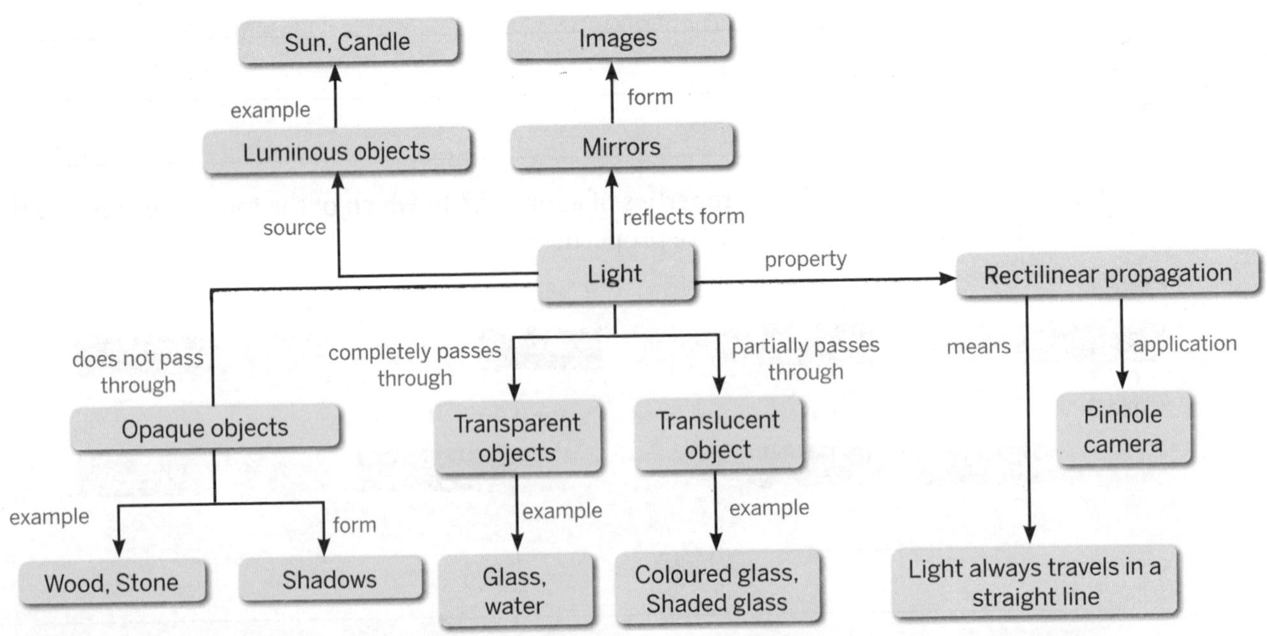

Worksheet 1

Name: ____________

Grade: A B C D

A. Choose the correct option.

1. The speed of light in vacuum is
   (a) 3 m/s (b) 3×10^8 km/s (c) 3,00,000 m/s (d) 3,00,000 km/s
2. Objects that give out light are called
   (a) sources of light (b) transparent objects
   (c) translucent objects (d) opaque objects
3. Which of the following is a non-luminous object?
   (a) Sun (b) Moon (c) Star (d) Firefly
4. Which among the following is an example for artificial source of light?
   (a) Star (b) Sun (c) Candle (d) Firefly
5. Butter paper and shaded glass are examples of ____________ materials.
   (a) transparent (b) translucent (c) luminous (d) opaque

B. Fill in the blanks using the correct words.

6. ____________ objects emit their own light. (Luminous/Non-luminous)
7. ____________ objects allow light to pass through them partially. (Transparent/Translucent)
8. A light ____________ is a single and very narrow path of light. (ray/beam)
9. ______________ is formed when light is blocked by an object. (A shadow/An image)
10. The colour of the shadow of an object is always ______________. (black/colour of the object)

C. Give one example for the following.

11. Artificial source of light ____________________
12. Luminous material ____________________
13. Transparent object ____________________
14. Translucent object ____________________
15. Opaque object ____________________

D. State whether the following statements are true (T) or false (F).

16. An image is formed when an opaque object is placed in the path of light.
17. Light can travel both in straight lines and in curved lines.
18. Shadow gives the two-dimensional picture of an object.
19. A mirror can form an image of an object in the absence of a screen.
20. Eclipses occur as a consequence of rectilinear propagation of light.

Worksheet 2

Name: ______

Grade: A B C D

A. Answer the following questions in a sentence.

1. What are natural sources of light?

______

2. What is meant by a light beam?

______

3. How can we change the shape of a shadow?

______

4. Write the factors responsible for the change in size of a shadow.

______

5. What will happen when light rays are blocked by an opaque object?

______

B. Answer the following questions in brief.

6. How will you differentiate a light ray and a light beam?

______

______

7. What is meant by rectilinear propagation of light?

______

______

8. Write a short note on shadows.

______

______

9. What is meant by opaque object? Give two examples.

______

______

10. Compare and contrast the length of your shadows formed during morning, noon and evening.

______

______

C. **Answer the following questions in detail.**

11. Write the differences between transparent and translucent materials. Give two examples for each.

12. What is meant by a shadow? What are the things necessary to form a shadow?

13. What is meant by sources of light? Differentiate between luminous and non-luminous objects by giving two examples of each.

14. Explain the conditions required for the formation of shadows.

15. Write a short note on the property of light that can be demonstrated by the given experimental setup. Suppose you move the screen B towards left. Will you still be able to see the light from the hole on the third cardboard?

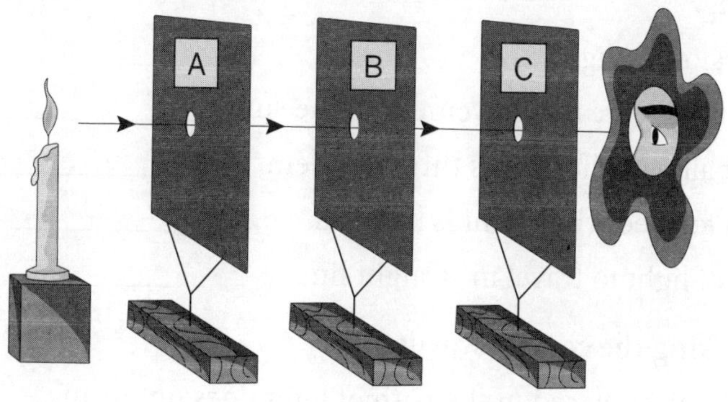

Worksheet 3

Name: ____________

Grade: (A) (B) (C) (D)

A. Choose the correct option.

1. A simple camera that consists of a lightproof box with a small hole on one side and a screen on the opposite side is called
   (a) pinhole camera (b) digital camera (c) OHP (d) LCD projector
2. Rough surfaces reflect light
   (a) evenly (b) unevenly (c) virtually (d) None of these
3. Identify the correct statement.
   (a) A pinhole camera has two lenses.
   (b) A pinhole camera does not have lenses.
   (c) A pinhole camera has only one lens.
   (d) The lens of a pinhole camera is same as that of a digital camera.
4. The image formed by a pinhole camera is
   (a) Real and inverted
   (b) Virtual and inverted
   (c) Real and upright
   (d) Virtual and upright
5. Images in the mirrors are formed due to________ of light.
   (a) reflection (b) absorption (c) blocking (d) None of these

B. State whether the following statements are true (T) or false (F)

6. The image formed by a pinhole camera is always inverted.
7. The size of the image of an object formed by a plane mirror is same as the size of the object.
8. The colour of the image of an object is always black.
9. The reflected rays of light from the object fall on our eyes, thereby making it visible to us.
10. In a pinhole camera, a bigger aperture lets in more light and we get a bright image.

C. Name the following.

11. Bouncing back of the light ____________
12. Photography to capture the movement of the sun ____________
13. Materials that allow light to pass through them ____________
14. The dark area formed when light is blocked ____________
15. The property of light to travel in straight line ____________

D. Fill in the blanks using the correct words.

16. A transparent object placed in the path of light does not form __________. (shadow/image)
17. An object is placed at a distance of 10 cm from a plane mirror. The distance of the image formed, measured from the mirror is __________ cm. (10/20)
18. The size of the image of an object formed by a plane mirror is __________ of the object. (same as the size/bigger than the size)
19. __________ is a device which makes use of the reflection of light by two parallel mirrors. (Periscope/Kaleidoscope)
20. Lunar eclipse occurs when the __________ of the Earth falls on the moon. (shadow/image)

Worksheet 4

Name:

Grade: A B C D

A. Answer the following questions in a sentence.

1. How are images formed?

2. What is the nature of reflected rays coming from a smooth surface?

3. What is the direction of the reflected rays when a light falls on a rough surface?

4. What is the colour of an image formed by a pinhole camera?

5. Why do we use a pinhole camera to view solar eclipse?

B. Answer the following questions in brief.

6. How are the objects in the nature visible to us?

7. What is meant by reflection of light?

8. Write an activity to show that the shadow of an object is affected by its distance from a light source.

9. What is meant by solargraphy?

10. Write any two characteristics of an image.

C. **Answer the following questions in detail.**

11. What is a pinhole camera?

12. Explain reflection of light with two examples.

13. What happens when the aperture size of a pinhole camera is increased?

14. How can we explain the role of rectilinear propagation in a pinhole camera?

15. Observe the pictures given below and explain the relationship between the size of the aperture and nature of image in a pinhole camera.

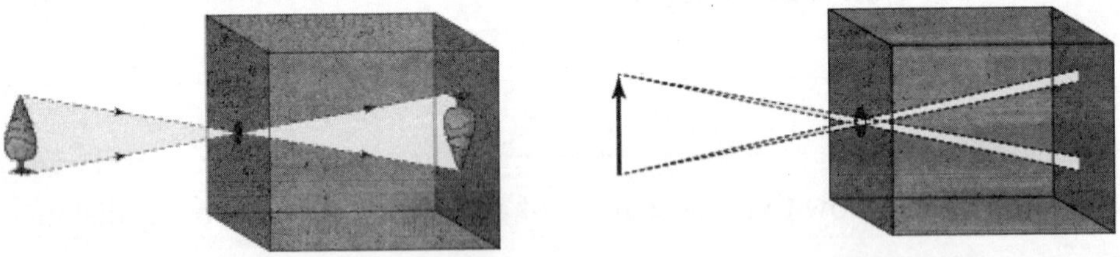

Image formation in a pinhole camera when the aperture is large

Image formation in a pinhole camera when the aperture is small

Worksheet 5

Name:

Grade: (A) (B) (C) (D)

A. Choose the correct option.

1. Which among the following is a luminous object?
   (a) Wood (b) Star (c) Earth (d) Moon
2. Which of the following terms is used for a single and very narrow path of light?
   (a) Light ray (b) Light beam (c) Pinhole camera (d) Opaque object
3. Which of the following terms is used for propagation of light in a straight line?
   (a) Reflection of light (b) Scattering of light
   (c) Solargraphy (d) Rectilinear propagation
4. Identify the correct sequence.
   (a) Light source → Screen → Object → Shadow
   (b) Screen → Light source → Object → Shadow
   (c) Light source → Object → Screen → Shadow
   (d) Object → Screen → Light source → Shadow
5. Pick the odd one out.
   (a) Glass (b) Wooden block (c) Cardboard sheet (d) Rubber

B. Match the following.

6. Transparent	(a) Wood
7. Translucent	(b) Sun
8. Opaque	(c) Shaded glass
9. Luminous material	(d) Torch
10. Artificial source of light	(e) Air

C. Give reasons for the following.

11. The shadows in the morning and evening are longer than the one in the afternoon.

    ____

12. The sound produced in a neighbouring classroom is audible to us, but candle lit is not visible.

    ____

13. In a completely dark room, we cannot see a reflection of ourselves in the mirror.

    ____

14. We can see the non-luminous objects even when they do not produce their own light.

    ____

15. Glass is the best material for a headlight cover of a vehicle.

    ____

D. Name the following.

16. Materials that do not allow light to pass through them ____________
17. The phenomenon due which an inverted image is formed in a pinhole camera ____________
18. The phenomenon responsible for the formation of an image ____________
19. An object which partially allows light to pass through it ____________
20. Surface that do not emit their own light ____________

E. Answer the following questions in detail.

21. Write an activity to show any one characteristic feature of a shadow.

____

____

____

22. Write the differences between a shadow and an image.

____

____

____

23. Differentiate between artificial and natural sources of light with suitable examples.

____

____

____

24. Explain an activity that shows light always travels in a straight line.

____

____

____

25. Write the differences between transparent and opaque materials by giving suitable examples.

____

____

____

15 Water

Recap of the Chapter

- Water is essential for the survival of all living things.
- Rain is the main source of water. Ponds, lakes, rivers, oceans and seas are other natural sources of water.
- Water is used for different purposes like performing bodily functions, irrigation, hydroelectric power generation, domestic use, industrial use, recreation and habitat.
- Water exists in three different forms—solid, liquid and gaseous.
- The states of water can be changed by heating and cooling.
- The continuous circulation of water from water bodies to the atmosphere and back is called water cycle.
- Water is circulated through different processes and at the end of each process; the state of the water is changed.
- Evaporation, transpiration, condensation, and precipitation are the different processes that are involved in a water cycle.
- A very long period of little or no rainfall results in drought.
- A condition where the ground gets submerged in water because of heavy rains and overflow of rivers is called flood.
- Conservation of water is essential for the present and the future generations.
- To reduce the usage of water, many ways such as drip irrigation, afforestation, reforestation and treatment of sewage, rainwater harvesting and industrial effluents were implemented.

Concept Map

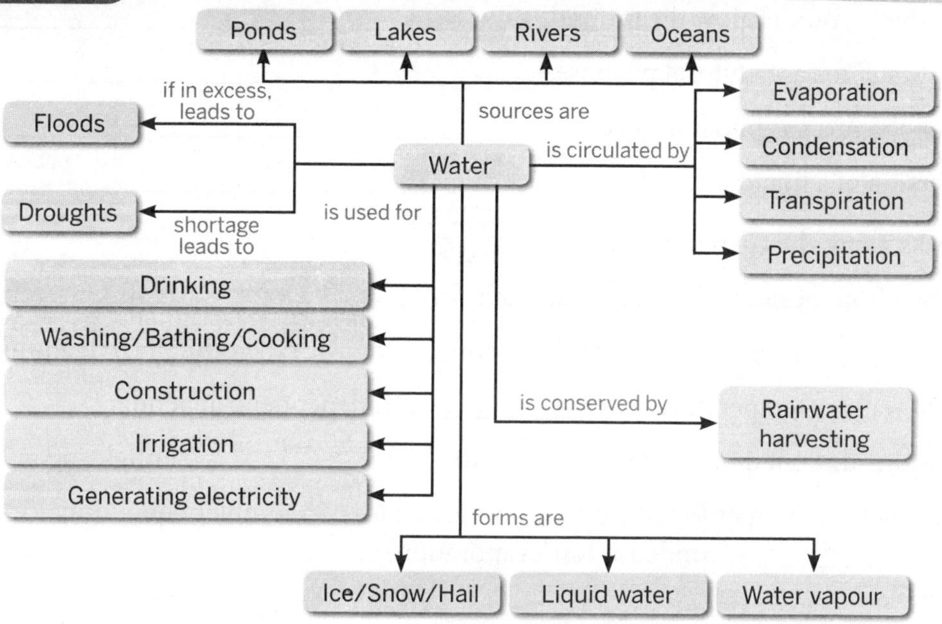

Worksheet 1

Name: ____________

Grade: (A) (B) (C) (D)

A. Choose the correct option.

1. Which of the following is not a source of fresh water?
   (a) Stream (b) Sea (c) River (d) Groundwater
2. Which of the following belongs to marine habitat?
   (a) Crocodiles (b) Crabs (c) Pelicans (d) Ducks
3. Which is the most commonly available state of water?
   (a) Solid (b) Liquid (c) Gaseous (d) Plasma
4. Through which of these processes does water vapour enter the air?
   (a) Precipitation (b) Condensation (c) Transpiration (d) Infiltration
5. Which of the following is a solid form of water?
   (a) Vapour (b) Steam (c) Hail (d) Fog

B. State whether the following statements are true (T) or false (F).

6. Only 3% of the total water on Earth is available for consumption.
7. Stored water from the dams and reservoirs are used to generate electricity.
8. Water exists in all the three states of matter.
9. Water circulates repeatedly in a circular manner to form water cycle.
10. Plants give out water vapour through their leaves due to evaporation.

C. Name the following.

11. The water found on the surface of the Earth ____________
12. The water that is found below the ground ____________
13. The use of water for agricultural purposes ____________
14. The conversion of a gas into liquid ____________
15. The conversion of a liquid into solid ____________

D. Fill in the blanks using the correct words.

16. Water drawn from ocean and sea is not suitable for ____________. (drinking/recreation)
17. ____________ of agricultural crops is essential along with the rainfall.(Irrigation/Harvesting)
18. The state of water is temperature ____________. (independent/dependent)
19. The wet clothes dry out quickly when exposed to ____________. (sunlight/neon light)
20. The process of formation of water droplets on the surface of a tumbler containing ice cubes is called ____________. (condensation/evaporation)

Worksheet 2

Name:

Grade:

Ⓐ Ⓑ Ⓒ Ⓓ

A. Answer the following questions in a sentence.

1. Name two domestic uses of water.

   ____

2. Name three solid forms of water seen in nature.

   ____

3. Define melting.

   ____

4. What are water droplets?

   ____

5. Define water cycle.

   ____

B. Answer the following questions in brief.

6. List some of the main uses of water.

   ____

   ____

7. How is hydroelectric power generated?

   ____

   ____

8. Write the uses of water in agriculture.

   ____

   ____

9. How can water be used for recreation?

   ____

   ____

10. What are the three states of water? Give one example for each.

    ____

    ____

C. **Answer the following questions in detail.**

11. Write an activity to show the effects of excess, deficient and sufficient amount of water on plants.

____

____

____

12. Explain an activity to show the process of evaporation and condensation.

____

____

____

13. Amount of water available on the Earth is constant. Comment.

____

____

____

14. Explain how water changes its state from one form to the other upon heating?

____

____

____

15. Label the given diagram.

Worksheet 3

Name:

Grade: A B C D

A. Choose the correct option.

1. Which of these states is prone to drought?
   (a) Assam (b) Rajasthan (c) Kerala (d) Odisha
2. What happens if there is no rain for a very long time?
   (a) Cyclone (b) Flood (c) Drought (d) Tsunami
3. The process of collection of rainwater in huge concrete pits is called
   (a) filtration of water (b) translocation of water
   (c) rainwater harvesting (d) infiltration
4. Which part of the plant helps in holding the soil during heavy rainfall?
   (a) Stem (b) Leaves (c) Flowers (d) Roots
5. Soil erosion leads to the removal of which layer of the soil?
   (a) Top layer (b) Middle layer (c) Bottom layer (d) None of these

B. State whether the following statements are true (T) or false (F).

6. Rainfall in India is even in all the regions.
7. The strong flow of water from the rivers damages the crops.
8. Water can be used liberally and it can be conserved.
9. Natural resources can be created.
10. Fixing the leakages can save a lot of water from being wasted.

C. Give one example for the following.

11. Drought-prone state ______________________
12. Flood-prone state ______________________
13. Water conservation at home ______________________
14. Water conservation in agricultural lands ______________________
15. Waterborne diseases ______________________

D. Fill in the blanks using the correct words.

16. Very poor or no rainfall leads to a condition called ____________. (flood/famine)
17. ____________ lead to epidemic diseases. (Floods/Drought)
18. ______________ is the liquid waste generated in an industrial plant or factory. (Effluent/Sludge)
19. Rainwater harvesting replenishes the ____________ levels.(surface water/groundwater)
20. Drip irrigation in agricultural lands can save a lot of ____________. (water/pesticides)

Worksheet 4

Name: ____________

Grade: Ⓐ Ⓑ Ⓒ Ⓓ

A. Answer the following questions in a sentence.

1. Write one effect of drought on plants.

   ____

2. What is famine?

   ____

3. What causes floods?

   ____

4. What do you mean by conservation?

   ____

5. What are effluents?

   ____

B. Answer the following questions in brief.

6. When do we say, there is a problem if water is not conserved?

   ____

   ____

7. What causes a drought-like condition?

   ____

   ____

8. How is plantation effective with the conservation of water?

   ____

   ____

9. Write a note on pollution of water caused due to sewage effluents.

   ____

   ____

10. What happens when it rains heavily?

    ____

    ____

C. **Answer the following questions in detail.**

11. What are the effects of drought?

12. What are the effects of flood?

13. What are the ways through which water wastage can be avoided at home?

14. Write a note on the condition you observe in the image given alongside.

15. Draw a neat-labelled diagram of rainwater harvesting.

Worksheet 5

Name:

Grade: A B C D

A. Choose the correct option.

1. Through which of the following water is brought back to the Earth in liquid state?
   (a) Vapours (b) Rain (c) Hail (d) Snow
2. Rise in water levels in the water bodies leads to which of these calamities?
   (a) Drought (b) Famine (c) Endemic (d) Floods
3. Which of the following is not a source of potable water?
   (a) Ocean (b) Lake (c) River (d) Well
4. Which of the following states is flood prone?
   (a) Kerala (b) Tamil Nadu (c) Odisha (d) Madhya Pradesh
5. Which of the following processes recharges the ground water?
   (a) Aquification (b) Transpiration (c) Infiltration (d) Transportation

B. State whether the following statements are true (T) or false (F).

6. The water in the sea has a high content of salt.
7. Water turns into ice on cooling.
8. Natural springs are the sources of water found in many lakes.
9. Groundwater is an example of potable water.
10. Plastic bottles, plates and cups need to be disposed in running water.

C. Give reasons for the following.

11. A huge amount of potable water is lost during distribution in main pipelines.

    ____

12. Shower in a bucket full of water aids in the conservation of water.

    ____

13. A wet floor dries after some time.

    ____

14. Water kept in a plate evaporates faster than the water kept in a glass.

    ____

15. Living things cannot survive without water.

    ____

D. Give two examples for the following.

16. Uses of water in recreation ______________________________
17. Uses of water in transportation ______________________________

18. Uses of water in bodily functions ____________________

19. Uses of water in industries ____________________

20. Uses of water in domestic purpose ____________________

E. Answer the following questions in detail.

21. Explain how oceans are important in a water cycle.

____

____

____

22. Define the different changes in the forms of water. Give one example for each.

____

____

____

23. Write a note on floods.

____

____

____

24. What are the different ways to conserve water?

____

____

____

25. Observe the given diagram and explain about the same.

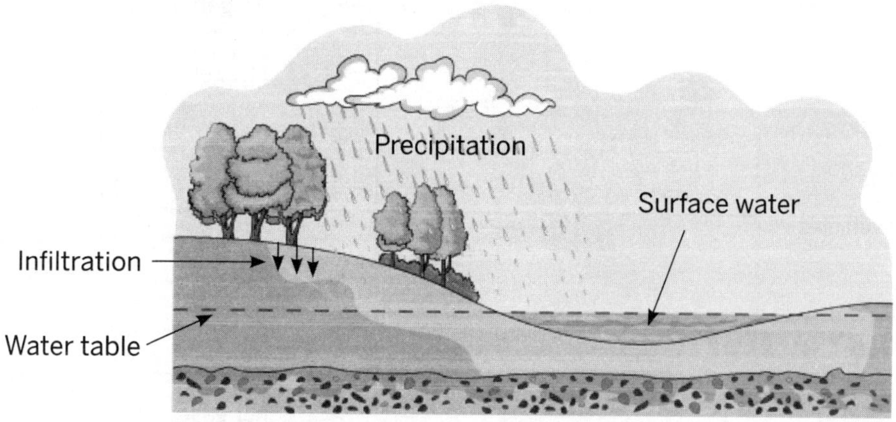

____

____

____

16 Air

Recap of the Chapter

- Earth is surrounded by a layer of air called atmosphere.
- Nitrogen, oxygen, carbon dioxide and other gases are present in the air.
- Along with the constituents of air, water vapour, dust and smoke are also present in the air.
- Nitrogen is the most abundant gas in the Earth's atmosphere. Oxygen is the second most abundant gas.
- Inhalation of dust, smoke and harmful gases cause respiratory diseases.
- The amount of water vapour in the atmosphere is called humidity. It varies from region to region.
- All living things breathe in oxygen and give out carbon dioxide during respiration.
- Different animals have different organs and parts of their body to help them breathe and respire.
- Plants breathe through stomata.
- Nature balances the amount of carbon dioxide and oxygen through various processes.
- Smoke, dust and other harmful gases in the air causes air pollution.
- Ozone layer that protects the Earth from UV radiation gets depleted by chemical pollutants in the air.

Concept Map

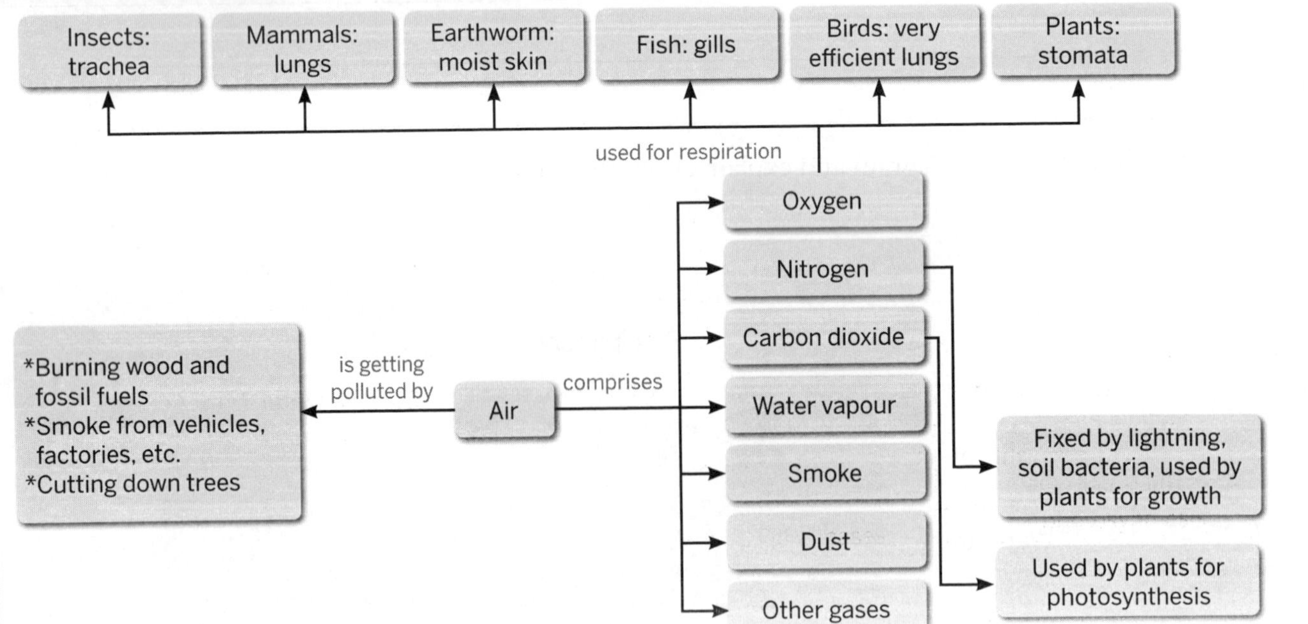

Worksheet 1

Name: ____________

Grade: A B C D

A. Choose the correct option.

1. 78% of the total constitution of atmosphere is composed of which of these gases?
   (a) Oxygen (b) Nitrogen (c) Carbon dioxide (d) Carbon monoxide
2. Which of these gases is required by all living organisms to survive?
   (a) Oxygen (b) Nitrogen (c) Carbon dioxide (d) Carbon monoxide
3. Which of these gases forms the basic building block of nucleic acid?
   (a) Oxygen (b) Nitrogen (c) Carbon dioxide (d) Carbon monoxide
4. Which of these organisms converts soil nitrogen into its simple form for the consumption of plants?
   (a) Earthworm (b) Virus (c) Insects (d) Bacteria
5. Which of these processes replenishes oxygen into the atmosphere?
   (a) Respiration (b) Excretion (c) Photosynthesis (d) Transpiration

B. State whether the following statements are true (T) or false (F).

6. Air is only present around us.
7. Air is a mixture of gases.
8. Oxygen is the most abundant gas present in the atmosphere.
9. Water vapour in the atmosphere is due to transpiration from plants.
10. The amount of water vapour present in the air is constant in all places.

C. Name the following.

11. The most abundant gas in the atmosphere ____________
12. The second most abundant gas in the atmosphere ____________
13. The gas released during respiration ____________
14. Inhalation of smoke and dust causes ____________
15. Atmospheric humidity is more during the season of ____________

D. Fill in the blanks using the correct words.

16. The layer of air surrounding the Earth is called ____________. (atmosphere/lithosphere)
17. Twenty-one per cent of the atmosphere consists of ____________. (carbon dioxide/oxygen)
18. Plants use ____________ during photosynthesis. (carbon dioxide/carbon monoxide)
19. Lotus plants have ____________ on the upper side of their leaves. (thorns/stomata)
20. The ____________ that living beings inhale is used to obtain energy from food they eat. (oxygen/nitrogen)

Worksheet 2

Name:

Grade: A B C D

A. Answer the following questions in a sentence.

1. How atmosphere is important to us?

2. What is atmospheric humidity?

3. Write one use of carbon dioxide.

4. What is respiration?

5. How does water enter the atmosphere?

B. Answer the following questions in brief.

6. List the other components that are found in the atmosphere along with the gases.

7. Write a note on dust and smoke present in the air.

8. How does carbon dioxide enter the atmosphere?

9. How can you find the dust present in the air inside a dark room?

10. List any three uses of air.

C. Answer the following questions in detail.

11. Write an activity to show that air is present around us.

12. Write an activity to show that water contains air.

13. Write the composition of air with the help of a diagram.

14. What is the role of carbon dioxide in extinguishing the fire?

15. What is required for the candle to burn continuously? Explain.

Worksheet 3

Name:

Grade: A B C D

A. Choose the correct option.

1. Which of the following organs is used by a cockroach for breathing?
   (a) Gills (b) Spiracles (c) Nose (d) Moist skin
2. Which gas is used by most living things for breathing?
   (a) Oxygen (b) Hydrogen (c) Nitrogen (d) Carbon dioxide
3. Which of the following is present in the stratosphere layer of our atmosphere?
   (a) Nitrogen (b) Water vapour (c) Ozone (d) Oxygen
4. Which of the following cannot be used by the plants directly from the atmosphere?
   (a) Oxygen (b) Carbon dioxide (c) Nitrogen (d) Water vapour
5. Which of the following chemicals depletes the ozone layer?
   (a) Chlorofluorocarbons (b) Hydrogen (c) Sulphur dioxide (d) Carbon dioxide

B. Match the following.

6. Human beings	(a) Gills
7. Grasshopper	(b) Moist skin
8. Earthworm	(c) Moist skin and lungs
9. Fish	(d) Nose
10. Frog	(e) Spiracles

C. Name the following.

11. The layer of atmosphere closest to the Earth ____________
12. The type of fuel that causes air pollution ____________
13. The medium of movement for airplanes and helicopters ____________
14. Tiny holes present on the underside of leaves to take in air ____________
15. The process through which plants give out oxygen ____________

D. State whether the following statements are true (T) or false (F).

16. Salamander cannot breathe underwater.
17. Air can be used to generate electricity.
18. Different organisms use different organs to breathe.
19. Ozone layer protects all living organisms on Earth.
20. Air provides medium for pollination of flowers.

Worksheet 4

Name:

Grade:

A B C D

A. Answer the following questions in a sentence.

1. Name the gases inhaled and exhaled by organisms while breathing?

____

2. Apart from oxygen and carbon dioxide, which other gas is used by plants and animals and how?

____

3. Name two aquatic plants that have stomata on the upper side of their leaves.

____

4. How do frogs and insects breathe?

____

5. Which hemisphere of the Earth possesses the most depleted layer of ozone? Why ozone is important to us?

____

B. Answer the following questions in brief.

6. Write a note on the respiratory diseases caused by air pollution.

____

____

7. Write a similarity between a human nose and stomata of a plant.

____

____

8. What is meant by humidity in the air? What is its effect?

____

____

9. Briefly describe the process of photosynthesis.

____

____

10. Describe the breathing organs of aquatic animals.

____

____

C. **Answer the following questions in detail.**

11. How do the bacteria present in the soil help in the growth of plants?

12. Describe the similarity in breathing organs of earthworms and frogs.

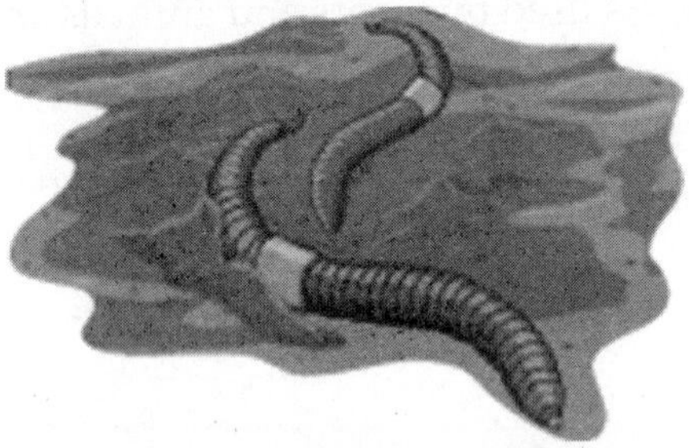

13. Write a note on the importance of air for all living things.

14. List the measures to reduce air pollution.

15. Describe the ozone layer.

Worksheet 5

Name: ____________

Grade: A B C D

A. Choose the correct option.

1. During respiration, which of the following gases is taken in by plants?
   (a) Oxygen (b) Carbon dioxide (c) Nitrogen (d) Carbon monoxide
2. Which of the following is converted into nitric acid by lightening and rain?
   (a) Nitrogen (b) Sulphur (c) Carbon dioxide (d) Oxygen
3. Which of the following terms is used for the blanket of air surrounding the Earth?
   (a) Atmosphere (b) Hydrosphere (c) Lithosphere (d) Biosphere
4. Which of the following increases carbon dioxide in the atmosphere?
   (a) Planting trees (b) Cutting down trees
   (c) Reducing vehicle exhaust (d) Cultivating aquatic plants
5. Which of the following animals uses lungs for breathing?
   (a) Whale (b) Monkey (c) Cow (d) All of these

B. Fill in the blanks using the correct words.

6. The chemical pollutants are responsible for depletion of __________ layer. (ozone/oxygen)
7. Air pollution can be controlled by reducing __________. (deforestation/afforestation)
8. The __________ cycle is influenced by temperature and movement of air. (water/oxygen)
9. Birds and insects fly in the medium of __________. (water/air)
10. Plants prepare their food by the process of __________. (respiration/photosynthesis)

C. Give reasons for the following.

11. Oxygen is required to release energy in living things.

    ____

12. The use of spray aerosols, refrigeration units and fire extinguishers must be reduced to prevent air pollution.

    ____

13. Burning of fossil fuels is bad for the environment.

    ____

14. Implying emission norm can improve the quality of air in cities.

    ____

15. Cities are more polluted than villages.

    ____

D. Complete the pattern.

16. Plants	:	Stomata	::	Fish	:	___________
17. Vehicle exhaust	:	___________	::	Oil spill	:	Water pollution
18. Refrigerants	:	Depletion of ozone	::	Deforestation	:	___________
19. Photosynthesis	:	Oxygen	::	Respiration	:	___________
20. Grasshopper	:	___________	::	Fish	:	Gills

E. Ans~~w~~er the following questions in detail.

21. How does increase in human activities contribute to air pollution?

____

____

____

22. Complete the given diagram and explain the balance of oxygen and carbon dioxide in nature.

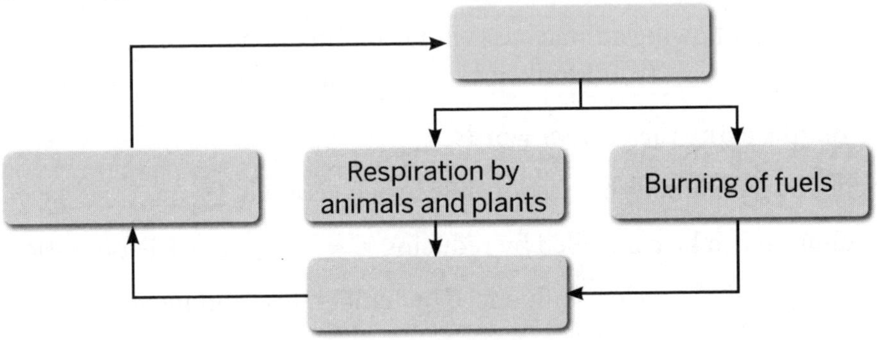

____

____

____

23. Aquatic animals live in water, but different animals have different breathing organ. Explain.

____

____

____

24. Differentiate between humidity of coastal and inland region.

____

____

____

25. Write an activity to show that air contains oxygen and nitrogen.

____

____

____

17 Waste Management

Recap of the Chapter

- Waste material is also called garbage.
- Wastes are of two main types, namely biodegradable and non-biodegradable.
- The biodegradable waste materials can be converted into compost and vermicompost.
- Compost materials act as natural fertilizers.
- The biodegradable waste is also added as landfills in low-lying areas.
- The 4R rule to conserve our environment is reuse, recycle, reduce and refuse.
- Municipalities are the authorities responsible for keeping the towns clean.
- Corporations are the authorities responsible for keeping the cities clean.
- *Panchayats* are responsible for keeping the villages clean.

Concept Map

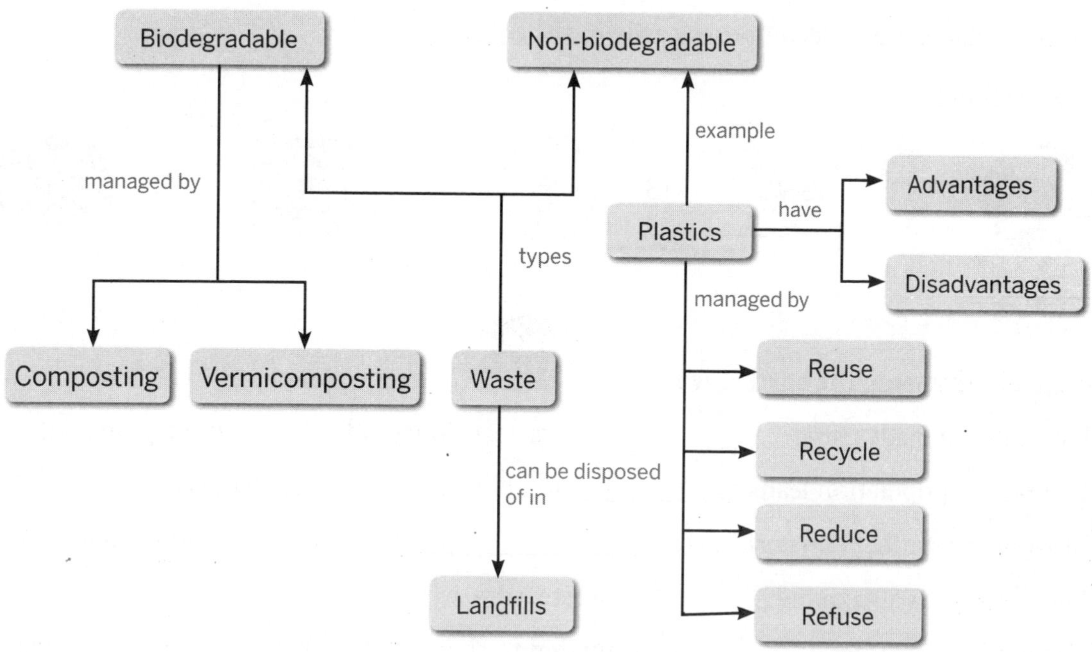

Worksheet 1

Name: ____________

Grade: A B C D

A. Choose the correct option.

1. Which of the following takes more time to decompose naturally?
   (a) Paper waste (b) Food waste (c) Twigs of plants (d) Animal skin
2. Which of the following cannot be recycled?
   (a) Biodegradable waste (b) Non-biodegradable waste
   (c) Kitchen waste (d) Plant and animal waste
3. Which of the following can be used to prepare compost?
   (a) Kitchen waste (b) Animal dung (c) Plant waste (d) All of these
4. Which of the following involves the use of a species of earthworms?
   (a) Vermicompost (b) Landfill
   (c) Animal waste (d) Trash
5. Garbage cannot be
   (a) reused (b) recycled
   (c) burnt (d) chemically degraded

B. State whether the following statements are true (T) or false (F).

6. The garbage collected at home is called trash.
7. Landfills can be converted into playground.
8. There are only disadvantages of using plastic.
9. Waste is grouped based on its property of decay.
10. Biodegradable waste decomposes easily in the soil.

C. Name the following.

11.  ____________

12.  ____________

13.  ____________

14.  ____________

15.  ____________

D. Fill in the blanks using the correct words.

16. Landfills are made using ____________ waste. (biodegradable/non-biodegradable)
17. Increase in population leads to ____________ in pollution. (increases/decreases)
18. Most of the synthetic materials are ____________. (biodegradable/non-biodegradable)
19. Plastic waste is also called ____________ waste. (dry/wet)
20. Vermicompost is ____________ as compared to chemical fertilizers. (expensive/inexpensive)

Worksheet 2

Name:

Grade: A B C D

A. Answer the following questions in a sentence.

1. What is the name of the process where waste materials are separated before disposal? What is its benefit?

____

2. What is the other name of kitchen waste and why is it called so?

____

3. What are recyclable materials also called?

____

4. What is the purpose of keeping two dustbins of different colours in the same place?

____

5. Which type of cans help us to keep our surroundings clean and how?

____

B. Answer the following questions in brief.

6. Why are reusable substances separated from garbage before adding them to landfills?

____

____

7. What is meant by organic manure?

____

____

8. What happens if the garbage is disposed off into water bodies?

____

____

9. How does burning the garbage contribute to air pollution?

____

____

10. What happens to the biodegradable waste disposed in a plastic bag?

____

____

C. **Answer the following questions in detail.**

11. Complete the picture given below by filling the blanks. Explain the process shown in the picture given below.

After decay ⟶ ____________

Kitchen waste + ____________ + Animal dung

____

____

____

12. Write a note on landfills.

____

____

____

13. Write a short note on different types of wastes generated in homes.

____

____

____

14. Explain the process of segregation of paper waste at home.

____

____

____

15. Describe the materials that are segregated as waste materials in schools.

____

____

____

Worksheet 3

Name:

Grade: A B C D

A. Choose the correct option.

1. Which of the following is the right way of discarding waste in public places?
   (a) Dustbins (b) Pathways (c) Water dispensers (d) Pots of plants
2. Which of the following can be reused?
   (a) Cloth bags (b) Polythene covers (c) Plastic boxes (d) All of these
3. Which of the following is a biodegradable waste that is kept in your school bag?
   (a) Notebook (b) Pen (c) Cotton napkin (d) Both (a) and (c)
4. Which of the following can be used as manure for plants in kitchen garden?
   (a) Vermicompost (b) Compost kitchen waste
   (c) Leaf litter (d) All of these
5. Which of the following must be thrown in the dustbins?
   (b) Chocolate wrapper (b) Ice-cream stick
   (c) Left-over food (d) All of these

B. State whether the following statements are true (T) or false (F).

6. People should not litter their surroundings.
7. Huge dustbins are placed in residential areas to dispose domestic waste.
8. Government authorities play a responsible role to keep our surroundings clean.
9. Old utensils at home can be recycled.
10. Aluminium utensils cannot be reused many a times.

C. Complete the pattern.

11. Municipality	:	Clean city	::	____________	:	Clean village
12. Polythene bag	:	Cloth bag	::	Plastic cup	:	____________
13. Less paper usage	:	More trees	::	Plastic ban in public place	:	____________
14. Chocolate wrapper	:	____________	::	Kitchen waste	:	Compost pit
15. Used mobile phone	:	E-waste	::	Used syringes	:	____________

D. Fill in the blanks using the correct words.

16. Landfills pollute soil and __________. (air/ground water)
17. Waste generated by hospitals are called ______________ waste. (biomedical/dry)
18. E-waste is generated by the use of ______________ goods. (electronic/agricultural)
19. Animal waste can be ______________. (non-biodegradable/biodegradable)
20. Plastic waste can be ______________ and reused. (remoulded/recoloured)

Worksheet 4

Name: ______

Grade: A B C D

A. Answer the following questions in a sentence.

1. Give one way of reusing plastic at home.

______

2. Which 4R rule is used in product packaging?

______

3. Which of the 4R rules is used for non-biodegradable materials?

______

4. Which of the 4R rules is used for the disposal of non-biodegradable materials?

______

5. Name the nutrient-rich manure obtained from kitchen waste and mention its use.

______

B. Answer the following questions in brief.

6. What is meant by plastic-free zone?

______

______

7. What is the benefit of refusing the use of unnecessary plastic packaging materials?

______

______

8. Though paper is biodegradable, its usage is recommended to be reduced. Why?

______

______

9. What is the best way to manage non-biodegradable waste?

______

______

10. What is meant by reuse of a material?

______

______

C. **Answer the following questions in detail.**

11. How can trees be saved by reducing the use of paper in our day-to-day life?

____

____

____

12. Write the different terms represented by 4Rs and give one example for each.

Terms	One example

13. List the functions of the city corporation authority.

____

____

____

14. Describe the meaning of going green.

____

____

____

15. Do you think medical waste contaminates groundwater? Give reason for your answer.

____

____

____

Worksheet 5

Name:

Grade: (A) (B) (C) (D)

A. Choose the correct option.

1. Which of the following terms is used for throwing waste in our neighbourhood?
   (a) Littering (b) Reuse (c) Recycle (d) Reduction of waste
2. Which of the following must be done to reduce plastic pollution?
   (a) Burn it (b) Bury it (c) Recycle it (d) Throw it
3. Which of the following cannot be reused in the kitchen?
   (a) Aluminium foil (b) Steel vessel
   (c) Glass vessel (d) Copper water bottle
4. Which of the following can be added to compost pit at home?
   (a) Animal dung (b) Fruit peels (c) Left-over food (d) All of these
5. Which of the following is suitable for a landfill?
   (a) Small hill (b) Low-lying area (c) River (d) Pond

B. Match the following.

6. Public places	(a) Compost pit
7. Leaf litter	(b) Plastic free zone
8. Cut plastic bottles	(c) Cloth bags
9. Reuse	(d) Animal feeders
10. Used paper	(e) Recycled paper

C. Give reasons for the following.

11. Refusing plastic bags at shops reduces plastic production and pollution.

    ____

12. The best way to manage non-biodegradable waste is to reuse the waste already generated.

    ____

13. Waste management is not a one-person job.

    ____

14. We should segregate waste into biodegradable, non-biodegradable and recyclable before disposing it of.

    ____

15. Used plastic bottles can be put to use as flower pots.

    ____

D. Give one example for the following.

16. Biodegradable clothing material ____________
17. Biodegradable stationery items used by students ____________
18. Non-biodegradable stationery items found inside pencil boxes ____________
19. A type of worm used in vermicomposting ____________
20. A part of the garbage picked back before adding into landfills ____________

E. Answer the following questions in detail.

21. How does disposing industrial and domestic waste into seawater harm our ecosystem?

____

____

____

22. Recognize the pictures given below and write any three differences between them.

____

____

____

23. Which other colour bins do you suggest to separate and dispose waste and why?

____

____

____

24. Differentiate between compost and vermicompost.

____

____

____

25. Write the differences between biodegradable and non-biodegradable waste.

____

____

____

Notes